Helical Laser Beams

Victor V. Kotlyar · Eugeny G. Abramochkin · Alexey A. Kovalev

Helical Laser Beams

Springer

Victor V. Kotlyar
Image Processing Systems Institute
NRC "Kurchatov Institute"
Samara, Russia

Alexey A. Kovalev
Samara National Research University
Samara, Russia

Eugeny G. Abramochkin
Samara Branch of Lebedev Physical Institute
Russian Academy of Sciences
Samara, Russia

ISBN 978-3-031-92216-9 ISBN 978-3-031-92217-6 (eBook)
https://doi.org/10.1007/978-3-031-92217-6

This Springer imprint is published by the registered company Springer Nature Switzerland AG
The registered company address is: Gewerbestrasse 11, 6330 Cham, Switzerland

Introduction

Helical laser beams are light fields that have singularity points in their phase distribution, i.e., points where the phase is undefined. Optical vortices also have screw dislocations in their wave front and the energy in such a beam propagates in a spiral (helix). There are isolated points of zero intensity in the intensity distribution of optical vortices. The topological charge (TC) is one of the main characteristics of optical vortices. It is an integer equal to the number of phase jumps by 2π during a movement along a circle of infinitely large radius in the beam cross-section. TC is positive if the phase increases during the movement counterclockwise, and TC is negative otherwise. This definition of topological charge was given by M. V. Berry and it shows that the cross-section of a helical beam can contain both a finite and an infinite number of singular points (local optical vortices), which can be located on the periphery of the laser beam in areas with almost zero intensity. Optical vortices can have only an integer or an undefined value of topological charge. The TC is conserved during propagation in free space, as is the orbital angular momentum (OAM) of helical beams. The orbital angular momentum density of conventional helical beams (Laguerre-Gaussian, Bessel-Gaussian) normalized to the beam power is equal to TC. Therefore, each photon in such a beam is said to have an OAM equal to the topological charge multiplied by Planck's constant. It is known how to find the OAM of an axial (or off-axial) superposition of vortex beams. But it was not known how to find the TC of such a superposition until now. In this book, topological charges for superposition (coaxial and non-coaxial) of Laguerre-Gaussian and Bessel-Gaussian beams are obtained.

The book also considers new helical beams that are not structurally stable, but are Fourier-invariant. These are double Laguerre-Gaussian and Bessel-Gaussian beams, squared Laguerre-Gaussian and Bessel-Gaussian beams, autofocusing Laguerre-Gaussian beams. New families of almost structurally invariant helical beams are also considered: generalized asymmetric Laguerre-Gaussian and Hermite-Gaussian beams, and several types of Bessel-Bessel-Gaussian beams. The book also discusses helical Ince-Gaussian beams, which are presented as series expansions in Hermite-Gaussian modes, and analytical expressions for the orbital angular momentum are obtained for them. For all these beams, explicit complex amplitudes were found at any distance from the initial plane.

The book also examines the superposition of optical vortices in the form of a geometric progression and finds the topological charge of such a superposition. In the initial plane, it turned out to be half-integer, but when the helical beam propagates in free space, the TC becomes integer.

The results presented in the monograph have been obtained during the research supported by the grant of the Russian Science Foundation 23-12-00236.

Contents

About the Authors

Victor V. Kotlyar is Head of the Laboratory at Image Processing Systems Institute of the National Research Center "Kurchatov Institute" and Professor of Computer Science at Samara National Research University. He received his MS, Ph.D., and DrSc degrees in Physics and Mathematics from Samara State University (1979), Saratov State University (1988), and Moscow Central Design Institute of Unique Instrumentation, the Russian Academy of Sciences (1992). He is a co-author of 400 scientific papers, 10 books, and seven inventions. His current interests are diffractive optics and singular optics. e-mail: kotlyar@ipsiras.ru

Eugeny G. Abramochkin graduated from Kuibyshev State University in 1984 with a degree in Mathematical Physics. Doctor of Physical and Mathematical Sciences (2006), Eugeny Abramochkin works as a leading researcher at the Samara branch of the Lebedev Physical Institute of the Russian Academy of Sciences. The list of scientific works includes about 60 articles. Scientific interests are related to complex analysis, theory of special functions, and equations of mathematical physics. e-mail: ega@fian.smr.ru

Alexey A. Kovalev graduated (2002) from Samara National Research University, majoring in Applied Mathematics. He received his Doctor in Physics and Maths degree in 2012. He is a senior researcher of Laser Measurements laboratory at Image Processing Systems Institute of the National Research Center "Kurchatov Institute" and a professor of Computer Science at Samara National Research University. He is a co-author of more than 270 scientific papers and four books. His research interests are mathematical diffraction theory and optical vortices. e-mail: alanko@ipsiras.ru

Chapter 1
Asymmetric Laser Beams

1.1 Generalized Asymmetric Hermite-Gaussian and Laguerre-Gaussian Beams

Laser beams whose complex amplitudes can be described by explicit analytical relationships that are exact solutions of the paraxial propagation equation have invariably been the focus of attention of optical researchers. The fact is that knowledge of the complex amplitude of the beams at any distance from the source plane enables the full set of their parameters to be predicted, including the intensity distribution, intensity moments, total power of the beam, topological charge, orbital angular momentum, and beam divergence. When propagated in free space, such beams are normally either structurally stable (retaining their intensity pattern up to a scale and rotation) or weakly changing, with their complex amplitude expressed by an elegant mathematical relationship. While such beams are abound, the most familiar are Hermite-Gaussian (HG) and Laguerre-Gaussian (LG) laser beams. For the first time, conventional HG beams were discussed in Ref. [1] and elegant HG beams were studied in Ref. [2]. Various forms of generalized HG beams have also been discussed in Refs. [3–7]. More specifically, in Ref. [3] the generalized HG beams were defined as conventional HG beams with a parameter, whereas in Ref. [7] the generalized beams were obtained as superposition of conventional HG beams propagating with the same phase velocity. Combined Hermite–Laguerre-Gaussian beams proposed in Ref. [8] were shown to transform into conventional HG beams or LG beams depending on the specific value of the parameter. A relationship between the HG and LG beams was established in Ref. [9] and vortex HG beams constructed as superposition of conventional HG beams with phase shifts were studied in Ref. [10]. For the first time, the LG beams were analyzed as laser cavity modes in Ref. [11] before being shown to propagate in an ABCD system [12]. Issues relating to the study of LG modes included their transformation and focusing with the aid of an axicon [13, 14] and their representation through Wigner functions [15]. Non-paraxial variants of the HG and LG beams have also been analyzed [16], as well as studying double-frequency

V. V. Kotlyar et al., *Helical Laser Beams*, https://doi.org/10.1007/978-3-031-92217-6_1

LG beams and their transformation into conventional LG beams [17]. Other types of beams studied, included asymmetric LG beams [18] and laser-aided intra-cavity shaping of asymmetric HG beams [19], which were then transformed into asymmetric LG beams using a mode converter [9]. A technique for generating vector laser beams with the aid of LG beams has been proposed [20]. Generation of LG beams in a wide frequency band by second harmonic generation has been reported [21], whereas conventional LG beams can be realized using a standard spatial light modulator (SLM) [22].

In this work, we discuss variants of generalized HG and LG beams, using which we generate asymmetric generalized HG and LG beams, with the former obtained in two different variants. Asymmetric Bessel-Gaussian and LG beams have been known to show high stability upon propagation in a turbulent medium [23, 24], prompting the exploration of other types of asymmetric HG and LG beams as promising candidates for monitoring the atmospheric turbulences.

1.1.1 *Generalized Hermite-Gaussian and Laguerre-Gaussian Beams*

The complex amplitude of conventional HG and LG beams is well known:

$$\begin{aligned} HG_{m,n}(x,y,z) &= \frac{Cw_0}{w(z)}\exp\left(-\frac{x^2+y^2}{w^2(z)} - ik\frac{x^2+y^2}{2R(z)} \right. \\ &\quad \left. -i(n+m+1)\arctan\left(\frac{z}{z_0}\right)\right) \\ &\quad \times H_m\left(\frac{\sqrt{2}x}{w(z)}\right)H_n\left(\frac{\sqrt{2}y}{w(z)}\right), \\ C &= \frac{1}{2^{(n+m)/2}}\sqrt{\frac{2}{\pi n!m!}}, \end{aligned} \tag{1.1}$$

$$\begin{aligned} LG_{p,l}(r,\varphi,z) &= \frac{C_1 w_0}{w(z)}\exp\left(-\frac{r^2}{w^2(z)} - ik\frac{r^2}{2R(z)} \right. \\ &\quad \left. -i(2p+|l|+1)\arctan\left(\frac{z}{z_0}\right)\right) \\ &\quad \times \left(\frac{\sqrt{2}r}{w(z)}\right)^{|l|} L_p^{|l|}\left(\frac{2r^2}{w^2(z)}\right)\exp(il\varphi), \\ C_1 &= \sqrt{\frac{2p!}{\pi(p+|l|)!}}. \end{aligned} \tag{1.2}$$

In (1.1) and (1.2), (x, y, z) and (r, φ, z) are, respectively, Cartesian and cylindrical coordinates, $z_0 = kw_0^2/2$ is the Rayleigh range, k is the wavenumber, w_0 is the waist radius of the Gaussian beam, m, n, p, l are integer numbers, $w(z) = w_0(1 + z/z_0)^{1/2}$, $R(z) = z\left(1 + z_0^2/z^2\right)$ is the beam radius at distance z from the waist and the radius of the Gaussian beam wavefront, and $H_m(\xi)$, $L_n^m(\xi)$ are Hermite polynomial and adjoined Laguerre polynomials.

An elegant form of the HG and LG beams is written as:

$$eHG_{m,n}(x, y, z) = \frac{1}{q^{(m+n+2)/2}(z)} \exp\left(-\frac{x^2 + y^2}{w_0^2 q(z)}\right) H_m\left(\frac{x}{w_0 q(z)}\right) H_n\left(\frac{y}{w_0 q(z)}\right), \tag{1.3}$$

$$eLG_{p,l}(r, \varphi, z) = \frac{1}{q^{p+1}(z)} \exp\left(-\frac{r^2}{w_0^2 q(z)} + il\varphi\right) \left(\frac{r}{w_0 q(z)}\right)^{|l|} L_p^{|l|}\left(\frac{r^2}{w_0^2 q(z)}\right), \tag{1.4}$$

where $q = 1 + iz/z_0$.

A generalized elegant version of HG and LG beams looks as follows:

$$gHG_{m,n}(x, y, z|a, b) = \frac{1}{q(z)} \exp\left(-\frac{x^2 + y^2}{w_0^2 q(z)}\right) \left(\frac{1 - aq(z)}{q(z)}\right)^{m/2} \times \left(\frac{1 - bq(z)}{q(z)}\right)^{n/2} H_m\left(\frac{x}{w_0\sqrt{q(z)(1 - aq(z))}}\right) H_n\left(\frac{y}{w_0\sqrt{q(z)(1 - bq(z))}}\right), \tag{1.5}$$

$$gLG_{n,m}(r, \varphi, z|a) = \frac{1}{q(z)} \exp\left(-\frac{r^2}{w_0^2 q(z)} + im\varphi\right) \times \left(\frac{1 - aq(z)}{q(z)}\right) \left(\frac{r}{w_0 q(z)}\right)^{|m|} L_n^{|m|}\left(\frac{r^2}{w_0^2 q(z)(1 - aq(z))}\right). \tag{1.6}$$

In (1.5) and (1.6), a and b are dimensionless parameters. Putting $a = b = 0$, the generalized beams in (1.5) and (1.6) is converted to the elegant beams in (1.3) and (1.4). If $a = b = 1/2$, generalized beams from Eqs. (1.5) and (1.6) reduce respectively to the conventional propagation-invariant HG and LG beams from Eqs. (1.1) and (1.2). Other forms of generalized Gaussian beams can also be found [5].

1.1.2 Asymmetric Generalized Hermite-Gaussian and Laguerre-Gaussian Beams

Asymmetric LG beams obtained by shifting the argument of the complex amplitude function into the complex domain have earlier been reported [18]. Laser-generated asymmetric LG beams that were then converted into asymmetric LG modes using a mode converter [9] have been realized [19]. The asymmetric LG modes have been used to steer microparticles along an arc [25] as well as for non-invasively capturing and moving biological microobjects [26]. We note that it is also possible to realize asymmetric HG beams from asymmetric LG beams using a mode converter [9]. In this work we demonstrate that asymmetric HG and LG beams are composed of a finite set of conventional generalized HG and LG beams of Eqs. (1.5) and (1.6). Actually, arguments of the Hermite and Laguerre polynomials can be shifted using the well-known [illegible] [27, 28]:

$$\sum_{s=0}^{n}\frac{n!}{s!(n-s)!}(2t)^{n-s}H_s(x)=H_n(x+t),$$

$$\sum_{s=0}^{n}\frac{t^s}{s!}L_{n-s}^{m+s}(x)=L_n^m(x-t). \tag{1.7}$$

Making use of the expressions in (1.7), the following relationships for complex amplitudes of LG and HG beams can be shown to hold:

$$\begin{aligned}
\sum_{s=0}^{n}\frac{t^s}{s!}gLG_{n-s,m+s}(r,z|a) &= \frac{1}{q(z)}\exp\left(-\frac{r^2}{w_0^2q(z)}\right)\\
&\times\sum_{s=0}^{n}\frac{t^s}{s!}\left(\frac{1-aq(z)}{q(z)}\right)^{n-s}\left(\frac{re^{i\varphi}}{w_0q(z)}\right)^{m+s}L_{n-s}^{m+s}\left(\frac{r^2}{w_0^2q(z)(1-aq(z))}\right)\\
&=\frac{1}{q(z)}\exp\left(-\frac{r^2}{w_0^2q(z)}\right)\left(\frac{1-aq(z)}{q(z)}\right)^n\left(\frac{re^{i\varphi}}{w_0q(z)}\right)^m\\
&\times L_n^m\left(\frac{r^2-tq(z)w_0re^{i\varphi}}{w_0^2q(z)(1-aq(z))}\right)=gLG_{n,m}(r,z|a,t).
\end{aligned} \tag{1.8}$$

$$\begin{aligned}
&\sum_{s=0}^{m}\sum_{p=0}^{n}\binom{m}{s}\binom{n}{p}(2\tau)^{m-s}(2t)^{n-p}gHG_{s,p}(x,y,z|a,b)\\
&=\frac{1}{q(z)}\exp\left(-\frac{x^2+y^2}{w_0^2q(z)}\right)\sum_{s=0}^{m}\binom{m}{s}(2\tau)^{m-s}\left(\frac{1-aq(z)}{q(z)}\right)^{s/2}\\
&H_s\left(\frac{x}{w_0\sqrt{q(z)}\sqrt{1-aq(z)}}\right)
\end{aligned}$$

$$
\begin{aligned}
&\times \sum_{p=0}^{n} \binom{n}{p} (2t)^{n-p} \left(\frac{1 - bq(z)}{q(z)} \right)^{p/2} H_p \left(\frac{y}{w_0 \sqrt{q(z)} \sqrt{1 - bq(z)}} \right) \\
&= \frac{1}{q(z)} \exp\left(-\frac{x^2 + y^2}{w_0^2 q(z)} \right) \left(\frac{1 - aq(z)}{q(z)} \right)^{m/2} \\
&\times \left(\frac{1 - bq(z)}{q(z)} \right)^{n/2} H_m \left(\frac{x + \tau w_0 q(z)}{w_0 \sqrt{q(z)} \sqrt{1 - aq(z)}} \right) \\
&H_n \left(\frac{y + t w_0 q(z)}{w_0 \sqrt{q(z)} \sqrt{1 - bq(z)}} \right) \\
&= gHG_{m,n}(x, y, z | a, b, \tau, t).
\end{aligned}
\tag{1.9}
$$

In (1.8), the parameter a defines the ratio of widths of the Gaussian function and Laguerre polynomial whereas the complex parameter t defines the degree of LG beam asymmetry. Actually, it is the absolute value of t that affects the asymmetry. As seen from Eq. (1.8), the argument of t only rotates the transverse pattern clockwise by the angle of arg t. In a similar way, in (1.9) the parameters a and b define the relative width of the Hermite polynomial along the x- and y-axes compared to the Gaussian function width, whereas the complex parameters τ and t define the degree of HG beam asymmetry along the x- and y-axes. We note that the asymmetric LG beam in (1.8) is different from the similar asymmetric beam discussed in Ref. [11], because in the latter case the argument was shifted into the complex domain for both the Gaussian beam and the Laguerre polynomials, unlike Eq. (1.8) in which the argument shift into the complex plane took place only for the Laguerre polynomial.

Using generalized HG beam in (1.5), it is possible to create elliptical vortex HG beams different from those discussed in Ref. [10]. As a starting point, we use the familiar formula for the summation of Hermite polynomials [27]:

$$
\sum_{s=0}^{n} \binom{n}{s} t^s H_{n-s}(x) H_s(y) = \left(1 + t^2\right)^{n/2} H_n \left(\frac{x + ty}{\sqrt{1 + t^2}} \right).
\tag{1.10}
$$

Using Eqs. (1.10) and (1.5) for superposition of generalized Hermite beams, the elliptic vortex HG beams can be derived in the form:

$$
\begin{aligned}
&\sum_{s=0}^{n} \binom{n}{s} t^s gHG_{n-s,s}(x, y, z | a, b) \\
&\quad = \left(1 + t^2\right)^{n/2} q^{-1}(z) \exp\left(-\frac{x^2 + y^2}{w_0^2 q(z)} \right) \left(\frac{1 - Tq(z)}{q(z)} \right)^{n/2} \\
&\quad H_n \left(\frac{x + ty}{w_0 \sqrt{1 + t^2}} \frac{1}{\sqrt{q(z)(1 - Tq(z))}} \right),
\end{aligned}
\tag{1.11}
$$

where $T = (a + bt^2)/(1 + t^2)$.

In (1.11), substituting $t = i\alpha$—where α is a real constant—we obtain an elliptical optical vortex $x + i\alpha y$ in the right-hand side in the argument of the Hermite polynomial. Hence, near every real root of the Hermite polynomial, which are found on the horizontal axis (a and b are real parameters), there is an optical vortex with the topological charge of +1 ($\alpha > 0$) or −1 ($\alpha < 0$). The modulus of the total topological charge of beam (1.11) equals n. At $\alpha = \pm 1$, the vortex Hermite beam in (1.11) is converted into a single-ring LG beam.

1.1.3 Non-Orthogonality and Power of Generalized Laguerre-Gaussian Beams

With the parameter a being complex, the LG beams in (1.6) are not mutually orthogonal. Considering that the dot product of two solutions of a paraxial equation remains unchanged at any z, non-orthogonality of the generalized LG beams can be proven in the source plane (at the Gaussian beam waist). Actually, it can be shown that in the source plane, the dot product of two generalized LG beams of (1.6) is given by

$$\begin{aligned}
&\int_0^\infty \int_0^{2\pi} gLG^*_{n,m}(r,\varphi) gLG_{p.q}(r,\varphi) r dr d\varphi \\
&= \frac{\pi w_0^2 (n+m)!}{2^{(n+m+p+1)} n!} \binom{n}{p} \delta_{m,q} \\
&\times (1-2a)^{n-p} \left(\text{Re}a - |a|^2\right)^p \\
&{}_2F_1\left(\begin{matrix} n+m+1, -p \\ n+1-p \end{matrix} \middle| -\frac{(1-2a)(1-2a^*)}{2(\text{Re}a - |a|^2)} \right),
\end{aligned} \tag{1.12}$$

where $\delta_{m,q}$ is the Kronecker symbol and ${}_2F_1(a, b; c\,|x)$ is a hypergeometric function. From (1.12), the generalized LG beams are seen to be orthogonal with respect to the azimuthal index because at $m \neq q$, the right-hand side of (1.12) vanishes. Meanwhile at different-valued radial indices, $n \neq p$, the right-hand side of (1.12) is non-zero. It is only at $a = 0$ that generalized LG beams are converted into conventional LG beams, becoming orthogonal with respect to both radial and azimuthal indices. Notably, from (1.12) follows an expression for the power W of a generalized LG beam:

$$W = \int_0^\infty \int_0^{2\pi} \left| gLG_{n,m}(r,\varphi) \right|^2 r dr d\varphi = \frac{\pi w_0^2 (n+m)!}{2^{(m+2n+1)} n!}$$

$$\times \sum_{s=0}^{n} 2^s \binom{n}{s} (1-2a)^{n-s} \left(1-2a^*\right)^{n-s}$$
$$\left(\text{Re}a - |a|^2\right)^s \binom{2n+m-s}{n+m}. \quad (1.13)$$

At $a = 0$, Eq. (1.13) is seen to describe the power of conventional LG beams:

$$W_{a=0} = 2^{2n} \int_0^\infty \int_0^{2\pi} \left|gLG_{n,m}(r,\varphi|a=0)\right|^2 r dr d\varphi$$
$$= \frac{\pi w_0^2 (n+m)!}{2^{(m+1)} n!}. \quad (1.14)$$

1.1.4 Topological Charge of Generalized Laguerre-Gaussian Beams and Their Anomalous Rotation Upon Propagation

For an arbitrary light field, its topological charge is defined by Berry's formula [29]:

$$TC = \frac{1}{2\pi} \lim_{r\to\infty} \int_0^{2\pi} \arg E(r,\varphi) d\varphi \quad (1.15)$$

with $E(r, \varphi)$ being the complex amplitude in the transverse plane and (r, φ) being the cylindrical coordinates in this plane.

The superposition (1.8) is described by a Laguerre polynomial of n-th degree. Thus, at $r \to \infty$, it is mostly determined by its highest-degree term and we can suppose that the right part of Eq. (1.8) reads as

$$gLG_{n,m}(r,z|a,t << r) = \frac{q^{-1}(z)}{n!} \exp\left(-\frac{r^2}{w_0^2 q(z)}\right) \left(\frac{1-aq(z)}{q(z)}\right)^n$$
$$\times \left(\frac{re^{i\varphi}}{w_0 q(z)}\right)^m \left(\frac{-r^2}{w_0^2 q(z)(1-aq(z))}\right)^n \quad (1.16)$$

As seen from Eq. (1.16), the TC of the whole superposition is thus equal to m. It is seen from Eq. (1.8) that if $m > 0$ then, upon propagation in free space, the transverse shape of the light beam should be rotated counterclockwise, as it typically happens for most light beams.

However, the asymmetry of the intensity distribution is affected only by the Laguerre polynomial, since the absolute values of all other multipliers in Eq. (1.8) are rotationally symmetric. For simplicity, we consider a simple case $a = 1/2$, since the denominator of the argument of the Laguerre polynomial becomes real-valued. The argument ξ is thus equal to

$$\xi = 2\frac{r^2 - tw(z)re^{i\varphi+i\eta}}{w^2(z)} \tag{1.17}$$

with $\eta = \arctan(z/z_0)$ being the Gouy phase and with $w(z) = w_0|q(z)|$. Expression (1.17) can be rewritten in the Cartesian coordinates:

$$\begin{aligned}\xi &= 2\frac{x^2+y^2}{w^2(z)} - 2t\frac{x+iy}{w(z)}e^{i\eta}\\ &= 2\left(\frac{x}{w(z)} - \frac{t\cos\eta}{2}\right)^2 + 2\left(\frac{y}{w(z)} + \frac{t\sin\eta}{2}\right)^2\\ &\quad - \frac{t^2}{2} - 2it\frac{y\cos\eta + x\sin\eta}{w(z)}.\end{aligned} \tag{1.18}$$

We see that the real part of the argument is, up to the constant $-t^2/2$, a distance from the point (x, y) to the point with coordinates

$$\begin{aligned}x_c &= \frac{t}{2}w(z)\cos\eta,\\ y_c &= -\frac{t}{2}w(z)\sin\eta.\end{aligned} \tag{1.19}$$

To determine the transformation of the beam transverse shape on propagation, we should choose some characteristic points whose evolution we are able to trace. Equation (1.18) indicates that we can choose zero-intensity points. Since all the roots of the Laguerre polynomials are real [30], then Im $\xi = 0$ in these points. Actually, these points reside on intersections of several concentric light rings (roots of the Laguerre polynomial) with the center given by Eq. (1.19), and of a straight line $y \cos\eta + x \sin\eta = 0$. In the initial plane ($z = 0$), the Gouy phase shift is zero ($\eta = 0$) and thus ξ is real at $y = 0$. The ring center is in the point $(t/2, 0)$. Thus, if $t > 0$, the considered zero-intensity points are to the right from the origin (Fig. 1.1a). Then, after propagation over some distance z: the rings center moves to the point given by Eq. (1.19), i.e. becomes rotated clockwise by the angle $-\eta$ with respect to the origin. The value ξ is real at $y = -x \tan\eta = -xz/z_0$, i.e. the straight line with real values ξ is also rotated clockwise by the angle η (Fig. 1.1b). In the far field, $z \gg z_0$ and $\eta = \pi/2$, the ring center is in the point $(0, -t/2)$ and ξ is real at $x = 0$ (Fig. 1.1c).

Thus, the beam (1.8), consisting of the LG beams with positive topological charges $m + s$, nevertheless, is rotated clockwise on propagation. Spiral beams with such

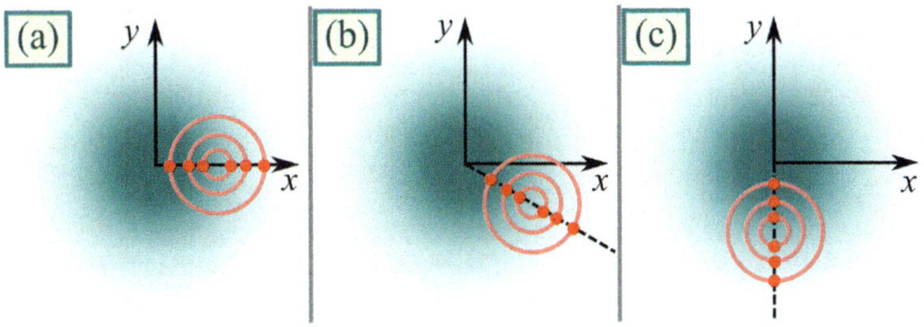

Fig. 1.1 Schematic evolution of the asymmetric generalized Laguerre-Gaussian beam from Eq. (1.8) shown as positions of zero-intensity points: in the initial plane (**a**), after propagation over some distance (**b**), in the far field (**c**). Blue color means the rotationally symmetric part of the beam (Gaussian envelope with an optical vortex), whereas pink rings denote the rings where the Laguerre polynomial is zero (if the imaginary part of its argument is neglected). Red dots thus denote the points where the beam intensity is zero

anomalous rotation have been investigated in [31], but the beams (1.8) are different since $2(n-s)+|m+s|+\theta(m+s)$ is not a constant at $\theta=-1$ and depends on s.

1.1.5 Numerical Modeling of Generalized Laguerre-Gaussian Beams

In this section, we describe the computation results of the asymmetric LG beams from Eq. (1.8). All distributions are obtained by two ways: by using the numerical Fresnel transform, implemented as a convolution with adopting the fast Fourier transform, and by the theoretical expression. All intensity distributions, obtained by these two ways, are visually indistinguishable, while the phase distributions are different only in low-intensity areas. This confirms the correctness of the formulae (1.8) for the complex amplitudes on space propagation.

Shown in Fig. 1.2 are intensity and phase distributions of the symmetric and asymmetric generalized Laguerre-Gaussian beams from Eq. (1.8) in several transverse planes.

As seen in Fig. 1.2, if $a = 1/2$ and $t = 0$, the asymmetric generalized Laguerre-Gaussian beam reduces to the conventional shape-invariant Laguerre-Gaussian beam. When $t = 0.25$, the beam becomes asymmetric and, as is predicted by the above theory, is rotated clockwise on propagation in space. We computed the topological charge in each plane. The obtained values are 3.0000 (Fig. 1.2m), 2.9981 (Fig. 1.2n), 2.9971 (Fig. 1.2o), 2.9982 (Fig. 1.2p), i.e. the theoretical value of 3 is confirmed.

Now we will see what happens when the asymmetry parameter is further increased. Figure 1.3 illustrates the intensity and phase distributions of the symmetric and asymmetric generalized Laguerre-Gaussian beams from Eq. (1.8) in several transverse planes when, instead of $t = 0$ and $t = 0.25$, we used the values $t = 1$ and $t = 2$.

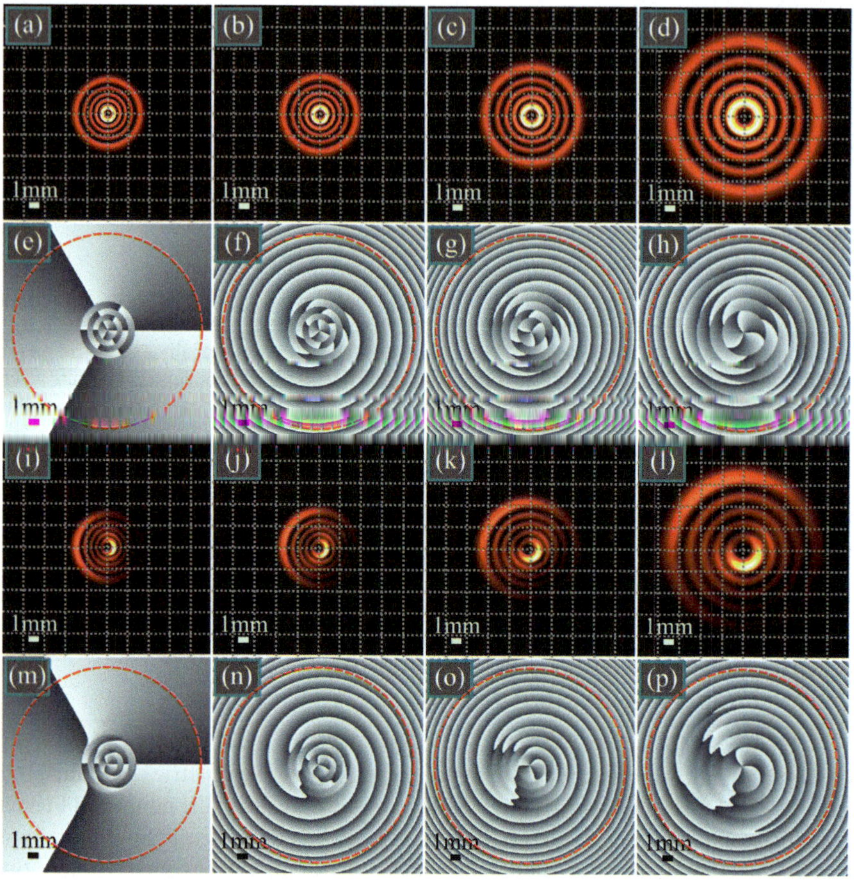

Fig. 1.2 Intensity (**a–d**, **i–l**) and phase (**e–h**, **m–p**) distributions of the asymmetric generalized Laguerre-Gaussian beams from Eq. (1.8) in several transverse planes for the following computation parameters: wavelength $\lambda = 532$ nm, Gaussian beam waist radius $w_0 = 1$ mm, lower index of the LG beam $n = 4$, upper index (topological charge) $m = 3$, scaling factor $a = 1/2$, asymmetry parameter $t = 0$ (**a–h**) and $t = 0.25$ (**i–p**), propagation distances $z = 0$ (**a**, **e**, **i**, **m**), $z = z_0/2$ (**b**, **f**, **j**, **n**), $z = z_0$ (**c**, **g**, **k**, **o**), $z = 2z_0$ (**d**, **h**, **l**, **p**). Dashed circles on the phase distributions denote the circles of the TC computation. Scale mark in each figure is 1 mm

According to Fig. 1.3, when the asymmetry parameter is increased, only the central and the outermost rings become bright while the other rings become pale. It is also seen that the intensity distribution tends to attain a shape of a single light ring (Fig. 1.3i–l). This is explainable since, when the asymmetry parameter t in Eq. (1.8) grows, the most significant weight in this superposition finally acquires the last LG beam, i.e. the beam with $s = n$. This is exactly a single-ringed LG beam (with the upper index of $m + n$).

The computed topological charge values in each plane are respectively as follows: 3.0000 (Fig. 1.3e), 2.9982 (Fig. 1.3f), 2.9972 (Fig. 1.3g), 2.9985 (Fig. 1.3h), 3.0000

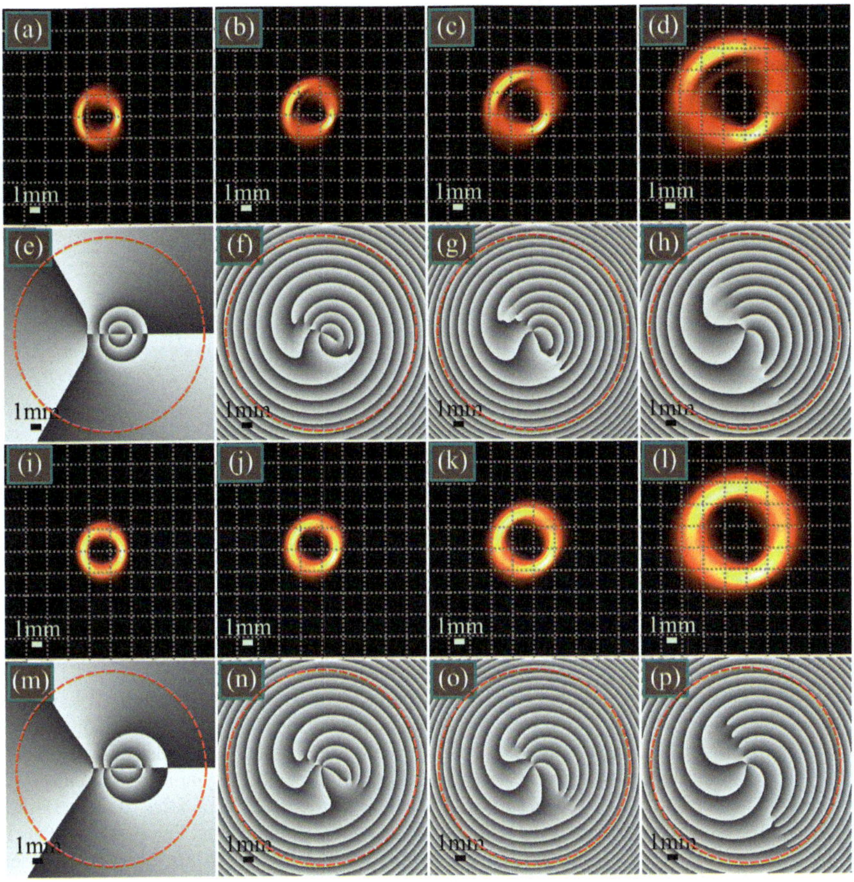

Fig. 1.3 Intensity (**a**–**d**, **i**–**l**) and phase (**e**–**h**, **m**–**p**) distributions of the asymmetric generalized Laguerre-Gaussian beams from Eq. (1.8) in several transverse planes for the following computation parameters: wavelength $\lambda = 532$ nm, Gaussian beam waist radius $w_0 = 1$ mm, lower index of the LG beam $n = 4$, upper index (topological charge) $m = 3$, scaling factor $a = 1/2$, asymmetryy parameter $t = 1$ (**a**–**h**) and $t = 2$ (**i**–**p**), propagation distances $z = 0$ (**a**, **e**, **i**, **m**), $z = z_0/2$ (**b**, **f**, **j**, **n**), $z = z_0$ (**c**, **g**, **k**, **o**), $z = 2z_0$ (**d**, **h**, **l**, **p**). Dashed circles on the phase distributions denote the circles of the TC computation. Scale mark in each figure is 1 mm

(Fig. 1.3m), 2.9982 (Fig. 1.3n), 2.9972 (Fig. 1.3o), 2.9988 (Fig. 1.3p), i.e. the theoretical value of 3 is confirmed.

When the asymmetry parameter t becomes large enough, the single-ringed LG beam completely overwhelms in the superposition and the intensity distribution looks like a homogeneous ring (Fig. 1.4). It may seem from Eq. (1.8) that this overwhelming leads to the change of the topological charge to $m + n$. However, Fig. 1.4 demonstrates that it remains equal to m, since there are additional n peripheral optical vortices with the charge of -1 (insets in Fig. 1.4i–l). For instance, when the topological charge is computed over a circle of radius 9 mm (circles in Fig. 1.4e–h), the obtained values are

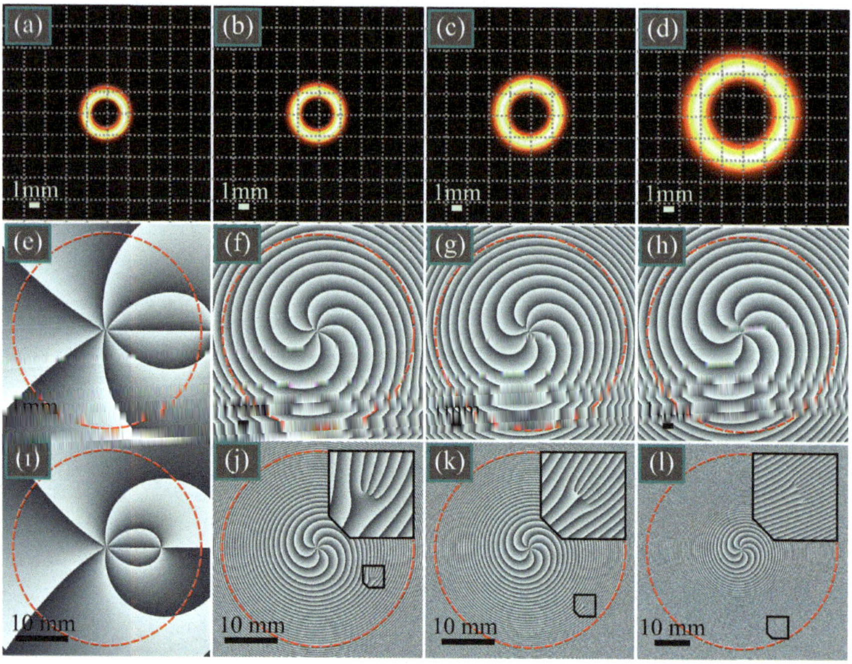

Fig. 1.4 Intensity (**a–d**) and phase (**e–l**) distributions of the asymmetric generalized Laguerre-Gaussian beams from Eq. (1.8) in several transverse planes for the following computation parameters: wavelength $\lambda = 532$ nm, Gaussian beam waist radius $w_0 = 1$ mm, lower index of the LG beam $n = 4$, upper index (topological charge) $m = 3$, scaling factor $a = 1/2$, asymmetryy parameter $t = 10$, propagation distances $z = 0$ (**a**, **e**, **i**), $z = z_0/2$ (**b**, **f**, **j**), $z = z_0$ (**c**, **g**, **k**), $z = 2z_0$ (**d**, **h**, **l**). Dashed circles on the phase distributions denote the circles of the TC computation. Phase distributions in the bottom row correspond to those from the middle row, but shown for larger areas. The insets (**i–l**) show the fourfold zoomed regions with additional vortices

7.0017 (Fig. 1.4e), 6.9980 (Fig. 1.4f), 6.9946 (Fig. 1.4g), 6.9968 (Fig. 1.4h). When the computation radius is increased twofold (Fig. 1.4i–k) or threefold (Fig. 1.4l), the obtained values are 3.0000 (Fig. 1.4i), 2.9704 (Fig. 1.4j), 2.9534 (Fig. 1.4k), 2.9625 (Fig. 1.4l).

1.1.6 Numerical Modeling of Generalized Hermite-Gaussian Beams

At first, we compare propagation-invariant and elegant HG beams. Figure 1.5 illustrates the intensity and phase distributions of the conventional HG beams from Eq. (1.1) whose shape is conserved upon propagation in both transverse directions (a–h), of elegant HG beams from Eq. (1.3) whose shape is changed, and of the hybrid generalized HG beams from Eq. (1.5) whose shape is conserved in one transverse direction and is changed in another direction.

As seen in Fig. 1.5, all three beams are symmetric with respect to the Cartesian axes. Now we will apply some asymmetry to the hybrid beam from Fig. 1.5q–x. Since the beam is different in different transverse directions along the axes x and y, for the changes to be more prominent, we apply a different asymmetry along these axes (Fig. 1.6).

As it was expected, nonzero parameter τ breaks the intensity asymmetry with respect to the axis y. Similarly, nonzero parameter t breaks the intensity asymmetry with respect to the axis x. When both the parameters t and τ are nonzero, both asymmetries become broken.

Summing up, we have derived novel exact solutions of the paraxial propagation equation [32]. Relationships have also been deduced for complex amplitudes that describe the propagation of generalized asymmetric HG and LG beams in (1.8) and (1.9). The derived relationships contain parameter sets, with an amplitude of a generalized asymmetric LG beam described by two complex parameters, namely, the scale parameter and the asymmetry degree parameter. The amplitude of a generalized asymmetric HG beam contains four parameters, a pair for each Cartesian axis. With the original generalized elegant HG (1.5) and LG (1.6) beams not retaining their intensity pattern upon free-space propagation, the derivative generalized asymmetric HG and LG beams also do not retain their intensity pattern in the course of their propagation evolution. The generalized LG beams of Eq. (1.9) are not orthogonal with regard to the radial indices, meaning that the asymmetric LG beams of Eq. (1.10) are also non-orthogonal with regard to the radial indices. For the generalized LG beams, we have proved that their topological charge is independent on the asymmetry and equals the upper index of the Laguerre polynomial. We have also found that the asymmetric LG beam is anomalously rotated upon propagation in free space, i.e. it is rotated clockwise when the vortex factor has a positive topological charge.

Another finding of this work, is a relationship for the complex amplitude of a family of generalized vortex HG beams in Eq. (1.11), which is characterized by three parameters, with two parameters defining the beam scale and a complex parameter defining the asymmetry degree of elliptical optical vortices of unit topological charge, with the vortices centered at zeros of a Hermite polynomial.

The proposed generalized asymmetrical laser beams can be realized using an SLM and utilized for data transmission through atmospheric turbulence. The scale parameters and the parameter controlling the asymmetry degree offer additional degrees of freedom enabling creating a beam least prone to distortion due to turbulence.

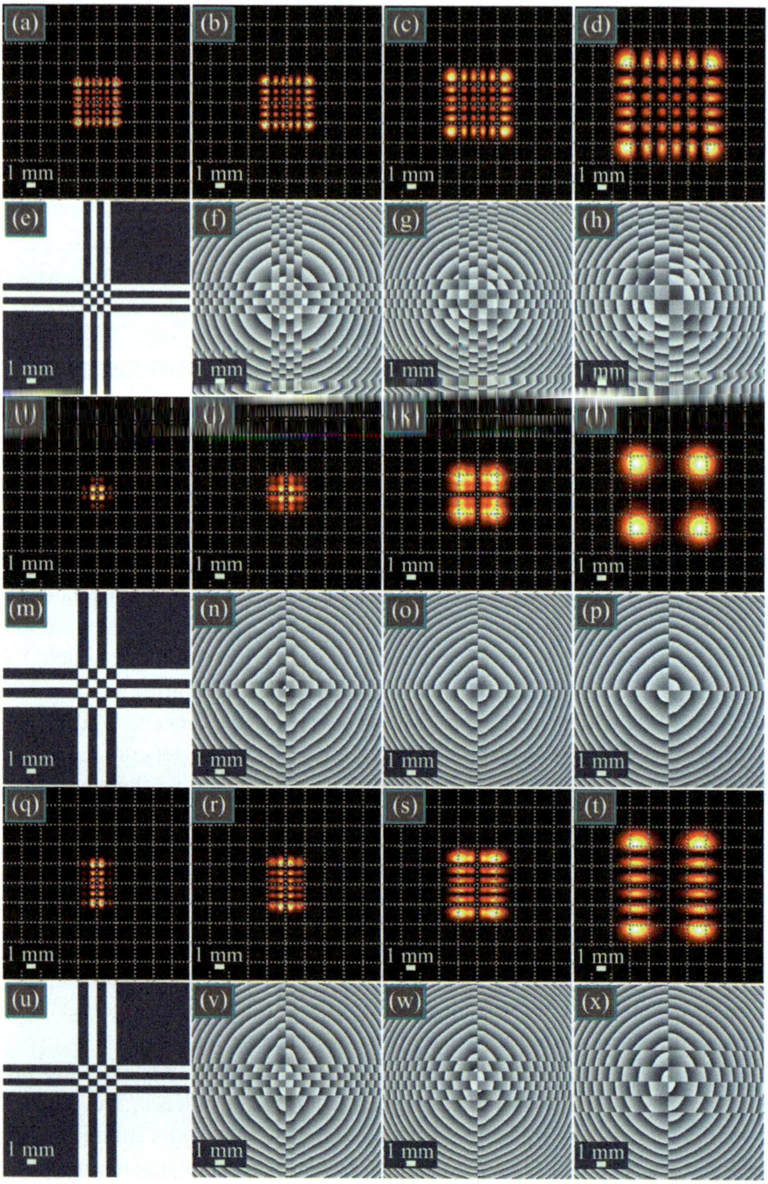

Fig. 1.5 Intensity (**a–d**, **i–l**, **q–t**) and phase (**e–h**, **m–p**, **u–x**) distributions of generalized Hermite-Gaussian beams from Eq. (1.5) (i.e. without the asymmetry) in several transverse planes for the following computation parameters: wavelength $\lambda = 532$ nm, Gaussian beam waist radius $w_0 =$ 1 mm, orders of the HG beams $m = 5$ and $n = 5$, scaling factors $a = 1/2$ and $b = 1/2$ (**a–h**), $a = 0$ and $b = 0$ (**i–p**), $a = 0$ and $b = 1/2$ (**q–x**), propagation distances $z = 0$ (**a**, **e**, **i**, **m**), $z = z_0/2$ (**b**, **f**, **j**, **n**), $z = z_0$ (**c**, **g**, **k**, **o**), $z = 2z_0$ (**d**, **h**, **l**, **p**). Scale mark in each figure is 1 mm

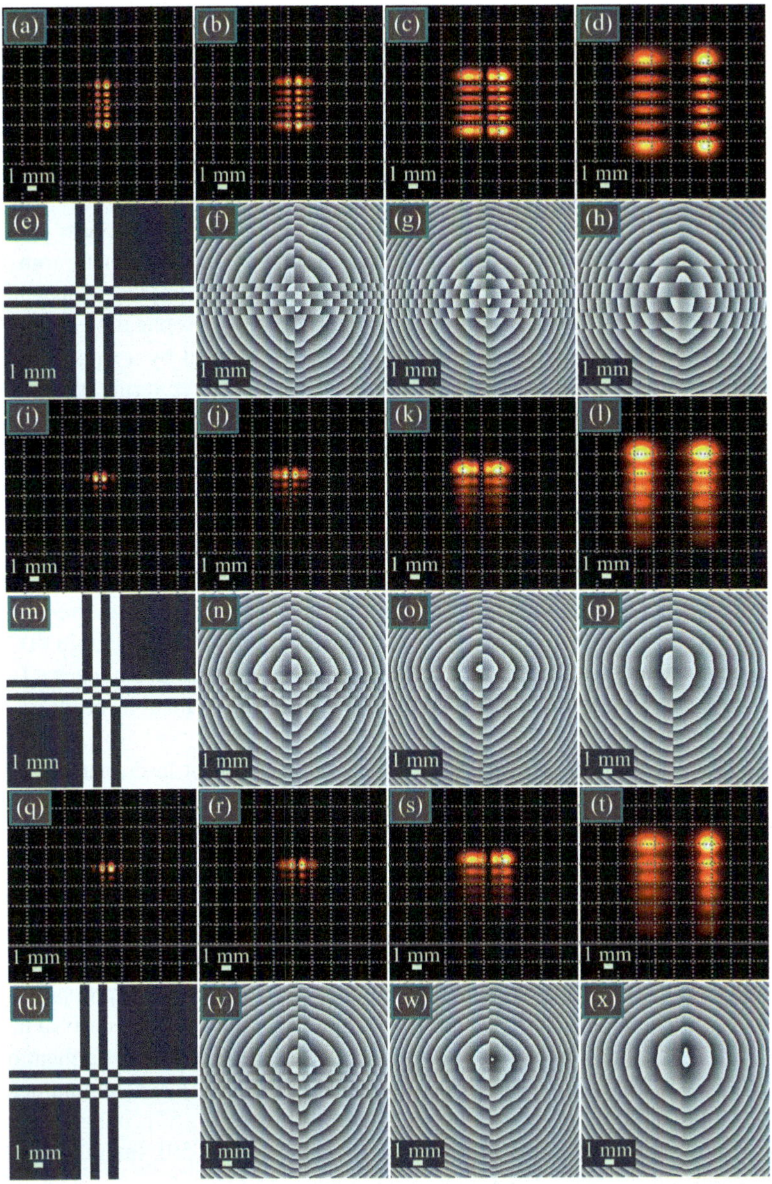

Fig. 1.6 Intensity (**a–d**, **i–l**, **q–t**) and phase (**e–h**, **m–p**, **u–x**) distributions of generalized asymmetric Hermite-Gaussian beams from Eq. (1.9) in several transverse planes for the following computation parameters: wavelength $\lambda = 532$ nm, Gaussian beam waist radius $w_0 = 1$ mm, orders of the HG beams $m = 5$ and $n = 5$, scaling factors $a = 0$ and $b = 1/2$, asymmetry parameters $t = 0$ and $\tau = 1/2$ (**a–h**), $t = 1/4$ and $\tau = 0$ (**i–p**), $t = 1/4$ and $\tau = 1/2$ (**q–x**), propagation distances $z = 0$ (**a**, **e**, **i**, **m**), $z = z_0/2$ (**b**, **f**, **j**, **n**), $z = z_0$ (**c**, **g**, **k**, **o**), $z = 2z_0$ (**d**, **h**, **l**, **p**). Scale mark in each figure is 1 mm

1.2 Topological Charge of Propagation-Invariant Laser Beams

Under propagation-invariant beams, we mean laser beams that conserve their intensity structure in the transverse section upon propagation in free space. An example of such beams is the Hermite–Laguerre-Gaussian (HLG) beams family. The HLG beams were first studied in [8] and found wide applications in modern optics. In [33], generalized HLG beams were obtained as well as their astigmatic transform. In [34], HLG beams were investigated in nonlinear medium. In [35], propagation of the HLG beams was studied in uniaxial crystals. In [36], propagation of the HLG beams was considered in a paraxial ABCD-system bounded by a square aperture. In [37], elegant HLG beams were obtained and the first three terms were given—nonparaxial corrections into the complex amplitude of such beams. In [38], a closed expression was given for incoherent HLG beams. In [38], the mutual intensity of the HLG beams was derived in the Shell model. In [39], the HLG beams were generated by using a metasurface. The HLG beams are a parametric beams family whose properties depend on the parameter. At a specific value of this parameter, the HLG beam reduces into the conventional Hermite-Gaussian (HG) beam. At another value of this parameter, the HLG beam reduces into the conventional Laguerre-Gaussian (LG) beam. It is known that the HG beams have zero topological charge (TC) and orbital angular momentum (OAM), whereas the LG beams have nonzero TC and OAM. However, it is still unknown which TC has the LG beam at an arbitrary value of the parameter.

The HLG beams are an example of propagation-invariant laser beams [40, 41]. Propagation-invariant beams can be represented as an axial superposition of the HG beams (or LG beams) with different weight amplitudes and phase delays. Mathematically, the complex amplitude of the HLG beams is equal to the sum of the HG beams with arbitrary complex coefficients. The sum of the indices of the Hermite polynomials in the HG beam should be constant. Then, all the HG beams in the superposition propagate with the same phase velocity and the transverse intensity distribution of the whole superposition change only in scale. The general expression for the OAM of a propagation-invariant superposition of the HG beams was derived in [42]. Propagation-invariant, including diffraction-free, beams found wide applications in optics [43, 44]. They are used in fundamental physics [45], telecommunications [46–48], optical cryptography [49–51], micromachines [52, 53], imaging systems [54–57], and for manipulating biological objects (cells) [58, 59]. Interesting studies were conducted in the works [60–62], studying the propagation-invariant beams, which are different from the HLG beams since each coefficient in the superposition of the HG beams was multiplied by a respective power of a complex number. At a certain value of this complex number, the OAM of such beams becomes zero, while at a small change of this complex number, the OAM changes abruptly. In the works [60–62], the expressions for the OAM were derived, but the TC of such generalized HLG beams was not considered in these works.

Obtaining the TC of propagation-invariant beams represented as a superposition of the HG beams with uncertain complex weights is generally a problem, equivalent by complexity to finding all zeros of a certain polynomial, given by its coefficients. The TC of such a superposition can be found for certain parametric laser beam families. In this work, we analytically derive the TC of the parametric family of the HLG beams at an arbitrary value of the parameter. In addition, we obtain the TC of several other propagation-invariant beams, for instance, two-parametric vortex Hermite-Gaussian beam, which is different from the single-parametric beam [10].

For convenience, we define here several terms important for this work. Vortices are screw wavefront dislocations of a laser beam. They are also called phase singularities. In the center of each vortex, the phase is undetermined and the intensity is therefore zero in these points. Thus, the vortices in the beam section can be detected by isolated intensity nulls. The topological charge of an optical vortex is an integer number of phase jumps by 2π along a closed contour of a large radius in the laser beam section. Thus, to obtain the TC, the phase distribution of the beam should be known. Orbital angular momentum density in each point of the beam section is defined as a product of conjugated complex amplitude $E^*(r, \varphi)$ and of its azimuthal angle derivative $dE(r, \varphi)/d\varphi$, where (r, φ) are the polar coordinates in the beam section. Physically, a nonzero OAM of a light beam means that the energy flow in the beam is propagating along a spiral.

1.2.1 *Propagation-Invariant Vortex Fields with Multiple Phase Singularities*

As was shown in [40, 41], a field given by

$$E_{\pm}(x, y, z) = \frac{1}{q}\exp\left(-\frac{x^2+y^2}{qw^2}\right)f\left(\frac{x \pm iy}{qw}\right), \tag{1.20}$$

where (x, y, z) are the Cartesian coordinates, w is the waist radius of the Gaussian beam, $q = 1 + iz/z_0$, $z_0 = \pi w^2/\lambda$ is the Rayleigh distance, λ is the wavelength, and $f(x \pm iy)$ is an arbitrary integer analytical function, is a solution to the paraxial Helmholtz equation (k is the wavenumber):

$$2ik\frac{\partial E}{\partial z} + \frac{\partial^2 E}{\partial x^2} + \frac{\partial^2 E}{\partial y^2} = 0. \tag{1.21}$$

Formula (1.20) describes propagation-invariant light fields with their intensity distribution conserving upon propagation in space (up to a scale and rotation around the optical axis). The rotation direction (clockwise or counter-clockwise) is defined by choosing the sign in the argument of the function f. Below, without loss of generality, we consider light fields with the sign "+". Formula (1.20) allows for an analytical description of light fields with arbitrary locations of optical vortices [40, 41] in the initial plane ($z = 0$) in the points with the Cartesian coordinates (a_k, b_k) ($k = 1$, …, M):

$$E(x, y, z) = \frac{1}{q}\exp\left(-\frac{x^2 + y^2}{qw^2}\right)\prod_{k=1}^{M}\left(\frac{x + iy}{q} - (a_k + ib_k)\right). \tag{1.22}$$

The topological charge (TC) of a light field can be evaluated with the Berry's formula [29]:

$$\mathrm{TC} = \frac{1}{2\pi}\lim_{r\to\infty}\mathrm{Im}\left\{\int_0^{2\pi}\frac{1}{E(r, \varphi, z)}\cdot\frac{\partial E(r, \varphi, z)}{\partial\varphi}d\varphi\right\}, \tag{1.23}$$

applying to the field complex amplitude in the polar coordinates $x = r\cos\varphi$, $y = r\sin\varphi$.

Formula (1.23) makes it possible to obtain the TC of an even more general beam, that instead of the Gaussian function in Eq. (1.22) is described by an arbitrary (even complex) radially symmetric multiplier $A(r)$. If the isolated intensity zeros of a propagation-invariant beam in the initial plane are in point with the coordinates $a_k + ib_k = r_k\exp(i\varphi_k)$, then the complex amplitude of such a beam is given by

$$E(r, \varphi, 0) = A(r)\prod_{k=1}^{M}\left(re^{i\varphi} - r_k e^{i\varphi_k}\right)^{m_k}, \tag{1.24}$$

where m_k are integer numbers equal to multiplicities of the intensity nulls.

Substituting Eq. (1.24) into Eq. (1.23), we obtain

$$\mathrm{TC} = \frac{1}{2\pi}\lim_{r\to\infty}\mathrm{Im}\left\{\int_0^{2\pi} ire^{i\varphi}\sum_{k=1}^{M}\frac{m_k}{re^{i\varphi} - r_k e^{i\varphi_k}}d\varphi\right\} = \sum_{k=1}^{M} m_k. \tag{1.25}$$

Thus, the TC of a propagation-invariant beam with M intensity zeros, each with a multiplicity m_k, is equal to the sum of these multiplicities. If the beam has optical vortices with both positive TCs and negative TCs, then its complex amplitude reads as

$$E(r, \varphi, 0) = A(r)\prod_{k=1}^{M}\left(re^{i\varphi} - r_k e^{i\varphi_k}\right)^{m_k}\prod_{\ell=1}^{N}\left(re^{-i\varphi} - r_\ell e^{-i\varphi_\ell}\right)^{n_\ell}. \tag{1.26}$$

Similarly to Eq. (1.25), the TC of the beam (1.26) can be obtained as

$$\mathrm{TC} = \sum_{k=1}^{M} m_k - \sum_{\ell=1}^{N} n_\ell. \tag{1.27}$$

The result (1.27) is almost obvious, since it means that the TC of a beam with M positive intensity zeros and with N negative zeros is equal to the algebraic sum of multiplicities of all the zeros in the beam section. We note that the beam (1.26) is no longer propagation-invariant.

1.2.2 Topological Charge of the Hermite–Laguerre-Gaussian Beams

In practice, the positions of the intensity zeros and their multiplicity are not known, and in the experiment they can be found only by generating a interference pattern with forks. Instead of explicitly defining a light field by its intensity zeros (1.22) or (1.24), a light field can be represented by polynomials or their linear combinations. Since the Hermite-Gaussian (HG) beams and the Laguerre-Gaussian (LG) beams are propagation-invariant beams themselves, then a large amount of propagation-invariant beams can be represented as a finite superposition of HG or LG beams, for instance, by HG beams with their indices yielding the constant Gouy phase. Thus, below we study paraxial beams with complex amplitude being equal to a finite sum of the HG beams with the indices, the sum of which is a constant:

$$\begin{aligned} E(x, y, 0) &= \sum_{k=0}^{N} c_k \mathrm{HG}_{N-k,k}(x, y) \\ &= \exp\left(-\frac{r^2}{w^2}\right) \sum_{k=0}^{N} c_k H_{N-k}\left(\frac{\sqrt{2}x}{w}\right) H_k\left(\frac{\sqrt{2}y}{w}\right). \end{aligned} \tag{1.28}$$

Here, c_k are arbitrary complex coefficients and $H_k(x)$ are the Hermite polynomials defined by the following recurrent relation:

$$H_{n+1}(x) = 2xH_n(x) - 2nH_{n-1}(x) \tag{1.29}$$

with the first two polynomials equal to $H_0(x) = 1$, $H_1(x) = 2x$.

Mathematically, the propagation invariance of the HG beams upon propagation can be expressed by via the Fresnel transform:

$$\begin{aligned} &\mathrm{HG}_{n,m}(x, y, z) \\ &= \frac{1}{\lambda i z} \int_{-\infty}^{\infty} \int_{-\infty}^{\infty} \exp\left[\frac{i\pi}{\lambda z}\left([x-\xi]^2 + [y-\eta]^2\right)\right] \end{aligned}$$

$$\begin{aligned}&\iint \ldots \mathrm{HG}_{n,m}(\xi,\eta)d\xi d\eta \\ &= \frac{1}{|q|}\exp\left[\frac{izr^2}{z_0 w^2|q|^2} - i(n+m+1)\arctan\left(\frac{z}{z_0}\right)\right] \\ &\quad \mathrm{HG}_{n,m}\left(\frac{x}{|q|},\frac{y}{|q|}\right),\end{aligned} \tag{1.30}$$

where $q = q(z) = 1+iz/z_0$, $z_0 = \pi w^2/\lambda$ is the Rayleigh distance, λ is the wavelength of light, (ξ, η) and (x, y) are the transverse Cartesian coordinates in the initial plane and at a distance z, z is the third Cartesian coordinate directed along the optical axis of the beam, and $(n + m + 1)\arctan(z/z_0)$ is the Gouy phase. As seen from Eq. (1.30), the HG beams with a constant sum of indices, i.e., with $n + m = N$, propagate with the same phase velocity, and therefore a superposition (1.28) of such beams is propagation-invariant. Below, we use dimensionless coordinates $x = x/w$, $y = y/w$ for brevity.

In a general case, if the complex coefficients c_k are arbitrary, the TC of the field (1.28) cannot be found. Below, we obtain the TC for two families of propagation-invariant fields: for vortex Hermite beams and for Hermite–Laguerre-Gaussian (HLG) beams.

We start with the latter ones. The complex amplitude of the HLG beams in the initial plane can be represented as a superposition of the HG beams [1]:

$$\mathrm{HLG}_{n,m}(x, y|\theta) = \sum_{k=0}^{n+m} i^k c_k^{(n,m)}(\theta)\mathrm{HG}_{n+m-k,k}(x, y). \tag{1.31}$$

The vertical line in the left part is a separator between the arguments (x, y) and the parameter θ, coefficients $c_k^{(n,m)}(\theta)$ are pure real trigonometric polynomials of θ and are expressed via the Jacobi polynomials:

$$c_k^{(n,m)}(\theta) = (\cos\theta)^{n-k}(\sin\theta)^{m-k}P_k^{(n-k,m-k)}(-\cos 2\theta). \tag{1.32}$$

The orbital angular momentum (OAM) of the beams (1.31) is known [8] and equals OAM $= (n - m)\sin 2\theta$, but the TC was not yet derived theoretically. It is seen that if $\theta = 0$ then, since $\sin 0 = 0$, only one term with $k = m$ remains in the expansion (1.31). In this case, $c_m^{(n,m)}(0) = (-1)^m$ and the HLG beam reduces to an HG beam:

$$\mathrm{HLG}_{n,m}(x, y|0) = (-i)^m\mathrm{HG}_{n,m}(x, y). \tag{1.33}$$

The topological charge of the HG beams is zero. It is easy to note that the HLG beam reduces to the HG beam also at other values of the parameter θ: $\pi/2$, π, $3\pi/2$, 2π, since in these cases also only one term remains in the sum (1.31). Thus, for these values of the parameter θ, the TC of the beam (1.31) is also equal to zero. However, at $\theta = \pi/4$, the HLG beam reduces to an LG beam:

$$\mathrm{HLG}_{n,m}(x,y|\pi/4) = \begin{cases} (-1)^m 2^n m! e^{-r^2} (re^{i\varphi})^{n-m} L_m^{n-m}(2r^2), & (n \geq m), \\ (-1)^n 2^m n! e^{-r^2} (re^{-i\varphi})^{m-n} L_n^{m-n}(2r^2), & (n \leq m). \end{cases} \tag{1.34}$$

with $L_k^{\alpha}(\cdot)$ being the associated Laguerre polynomial. Evidently, the TC of the HLG beam in this case is equal to TC $= n - m$. Thus, a question arises: if the TC of the beam (1.31) is equal to zero at $\theta = 0$, $\pi/2$, and is equal to $n - m$ at $\theta = \pi/4$, then what is the TC value at other values θ from the interval $(0, \pi/2)$? For obtaining the TC of the HLG beams in the general case, we use the generating function of the coefficients $c_k^{(n,m)}(\theta)$ [8]:

$$\sum_{k=0}^{n+m} i^k c_k^{(n,m)}(\theta) x^{n+m-k} y^k = (x\cos\theta + iy\sin\theta)^n (x\sin\theta - iy\cos\theta)^m. \tag{1.35}$$

According to Eq. (1.23), the TC of a light field is obtained for the values x and y tending to infinity. Using an asymptotic expression of the Hermite polynomials, $H_n(x \gg 1) \approx (2x)^n$, we obtain for the beam (1.31):

$$\begin{aligned} \mathrm{HLG}_{n,m}(x \gg 1, y \gg 1|\theta) &= exp(-r^2) \sum_{k=0}^{n+m} i^k c_k^{(n,m)}(\theta) H_{n+m-k}\left(\sqrt{2}x\right) H_k\left(\sqrt{2}y\right) \\ &\approx exp(-r^2) \sum_{k=0}^{n+m} i^k c_k^{(n,m)}(\theta) (2\sqrt{2}x)^{n+m-k} (2\sqrt{2}y)^k \\ &= exp(-r^2)(2\sqrt{2})^{n+m} (x\cos\theta + iy\sin\theta)^n \\ &\quad (x\sin\theta - iy\cos\theta)^m. \end{aligned} \tag{1.36}$$

It is seen from Eq. (1.36) that two elliptic optical vortices with the topological charges n and $-m$ interact in the origin ($x = y = 0$). Thus, according to Eq. (1.27), the TC of the whole field equals $n - m$. Therefore, the TC of the beam (1.31) is equal to $n - m$ at an arbitrary value θ from the interval $(0, \pi/2)$. The right part of the Formula (1.36) reveals that if θ is equal to $\pi/2$, π, $3\pi/2$, or 2π, then the optical vortices disappear. In addition, if $m = n$ and $\theta = \pi/4$, the central optical vortex in the origin is also absent:

$$\begin{aligned} &(x\cos\theta + iy\sin\theta)^n (x\sin\theta - iy\cos\theta)^m \Big|_{\substack{m=n \\ \theta=\pi/4}} \\ &= 2^{-n}(x+iy)^n (x-iy)^n = 2^{-n}(x^2+y^2)^n. \end{aligned} \tag{1.37}$$

From Eq. (1.36) follows that, in the second quarter of the angle θ variation period, i.e., in the interval $(\pi/2, \pi)$, there are optical vortices with the topological charges $-n$ and m in the origin. Thus, the TC of the beam (1.31) is equal to $-(n - m)$, i.e., changing the parameter θ makes it possible to change the TC sign of the HLG beam. It is interesting to compare the OAM of the HLG beam, OAM $= (n - m)\sin 2\theta$, and its TC, TC $= n - m$. It is seen that the TC is independent of the parameter θ while it is, for instance, in the interval $(0, \pi/4)$, whereas the OAM for different θ changes from zero to maximum. We note that multiplying all the coefficients (1.31) by a respective power of a real-valued parameter α, $C_k\prime = a^k C_k$, can only change the sign of the TC of the HLG beam, since in this case we obtain instead of Eq. (1.37):

$$(x\cos\theta + i\alpha y\sin\theta)^n(x\sin\theta - i\alpha y\cos\theta)^m. \tag{1.38}$$

1.2.3 Topological Charge of a Two-Parametric Vortex Hermite Beam

Propagation-invariant vortex Hermite beams were introduced and investigated in [33]. They are yet another example of the beam (1.28), whose coefficients allow the sum (1.28) to be written in a closed form and thus the TC can be obtained. To do this, we use the summation formula of the Hermite polynomials [27]:

$$\sum_{k=0}^{n+m} \frac{t^{n+m-k}\tau^k}{(n+m-k)!k!} H_{n+m-k}(x)H_k(y) = \frac{\left(t^2+\tau^2\right)^{(n+m)/2}}{(n+m)!} H_{n+m}\left(\frac{tx+\tau y}{\sqrt{t^2+\tau^2}}\right). \tag{1.39}$$

Then, for applying the summation Formula (1.39) for simplifying the sum (1.28), the coefficients in Eq. (1.28) should be chosen as follows:

$$C_k(t,\tau) = \frac{t^{n+m-k}\tau^k}{(n+m-k)!k!}. \tag{1.40}$$

Thus, the beam (1.28) with the coefficients (1.40) reads as

$$\begin{aligned} E_{n+m}(x,y) &= \exp\left(-x^2-y^2\right)\sum_{p=0}^{N} \frac{t^{n+m-k}(i\tau)^k}{(n+m-k)!k!} H_{n+m-k}\left(\sqrt{2}x\right)H_k\left(\sqrt{2}y\right) \\ &= \exp\left(-x^2-y^2\right)\frac{\left(t^2-\tau^2\right)^{(n+m)/2}}{(n+m)!} H_{n+m}\left(\sqrt{2}\frac{tx+i\tau y}{\sqrt{t^2-\tau^2}}\right) \end{aligned} \tag{1.41}$$

According to Eq. (1.41), if the beam parameters t and τ are real and not equal to each other, then all zeros of the Hermite polynomial, each being real-valued and residing on the horizontal axis $y = 0$, generate elliptic optical vortices (screw dislocations) with a unit TC if $t\tau > 0$ and with the TC of -1 if $t\tau < 0$. The topological charge of the whole beam (1.41) is equal to the number of roots, i.e., $n + m$ if $t\tau > 0$ and $-(n + m)$ if $t\tau < 0$. Thus, in the second case of Eq. (1.41), the TC of a propagation-invariant beam (1.28) is greater by magnitude than the TC of the beam (1.31). If $t = \tau$, the vortex Hermite beam (1.41) reduces to a single-ringed Laguerre-Gaussian beam and all elliptic optical vortices merge into a single standard circular optical vortex at the origin, whose topological charge is still equal either to $n + m$ or to $-(n + m)$. It is interesting that the TC of the beam (1.41) is not changed with changing parameters if these parameters t and τ do not change their sign.

1.2.4 Simple Optical Vortices

If we substitute the following binomial coefficients

$$C_k = \frac{(n+m)!}{(n+m-k)!k!}, \tag{1.42}$$

into Eq. (1.28), then we derive an asymptotic expression for large values of the arguments, similar to the sum (1.36):

$$\begin{aligned} E_{n+m}(x >> 1, y >> 1) &= e^{-(x^2+y^2)} \sum_{k=0}^{n+m} i^k \frac{(n+m)!}{k!(n+m-k)!} \\ &\times H_{n+m-k}\left(\sqrt{2}x\right) H_k\left(\sqrt{2}y\right) \\ &\approx e^{-(x^2+y^2)} (2\sqrt{2})^{n+m} \\ &\times \sum_{k=0}^{n+m} i^k \frac{(n+m)!}{k!(n+m-k)!} x^{n+m-k} y^k \\ &= e^{-(x^2+y^2)} (2\sqrt{2})^{n+m} (x+iy)^{n+m}. \end{aligned} \tag{1.43}$$

As seen from Eq. (1.43), the asymptotic argument is equal to $(n + m)\varphi$, where φ is the polar angle in the beam cross-section. Therefore, the TC of the field (1.28) with the coefficients (1.42) equals $n + m$. For deriving Eq. (1.43), the Newton's binomial theorem [27] was used:

$$\sum_{k=0}^{n} \frac{n!}{k!(n-k)!} x^k = (1+x)^n. \tag{1.44}$$

Another example of determining the topological charge of a superposition of optical vortices can be given by using the following finite sum [63]:

$$\sum_{k=0}^{n} \frac{1}{(2k)![(n-k)!]^2} x^k = \frac{(x-4)^n}{(2n)!} P_{2n}\left(\sqrt{\frac{x}{x-4}}\right), \tag{1.45}$$

where $P_{2n}(x)$ are the Legendre polynomials, orthogonal in the interval $[-1, 1]$, and having n simple real-valued zeros in this interval. Then, the asymptotic expression for large argument values for the beam (1.28) is given by

$$\begin{aligned} E_{n+m}(x >> 1, y >> 1) &= e^{-(x^2+y^2)} \sum_{k=0}^{n+m} i^k \frac{1}{(2k)![(n+m-k)!]^2} \\ &\times H_{n+m-k}\left(\sqrt{2}x\right) H_k\left(\sqrt{2}y\right) \\ &\approx e^{-(x^2+y^2)} (2\sqrt{2})^{n+m} \\ &\times \sum_{k=0}^{n+m} i^k \frac{1}{(2k)![(n+m-k)!]^2} x^{n+m-k} y^k \\ &= e^{-(x^2+y^2)} (2\sqrt{2})^{n+m} \\ &\times \frac{(-4x+iy)^{n+m}}{(2(n+m))!} P_{2(n+m)}\left(\sqrt{\frac{y(y-ix)}{y^2+4x^2}}\right). \end{aligned} \tag{1.46}$$

For deriving Eq. (1.46), the summation coefficients in Eq. (1.28) were chosen as follows:

$$C_k = \frac{1}{(2k)![(n+m-k)!]^2}. \tag{1.47}$$

According to Eq. (1.46), the TC of the beam (1.28) with the coefficients (1.47) is equal to $-(n+m)$, since there is an elliptic optical vortex in the center of the beam section, whose amplitude is proportional to an expression, $(-4x+iy)^{n+m}$. The Legendre polynomial does not have additional zeros since the real value of the Legendre polynomial argument in Eq. (1.46) equals 1. We note that if the coefficients (1.42) and (1.47) are multiplied by a real-valued parameter α, i.e., $C_k' = a^k C_k$, then such change in the coefficients changes the intensity distribution of the propagation-invariant beam (1.28) and changes its OAM, but does not change its TC.

1.2.5 Propagation-Invariant Beams in the Laguerre-Gaussian Basis

Instead of the HG basis, the complex amplitude of propagation-invariant light fields can be written in the Laguerre-Gaussian (LG) basis:

$$E_{\pm}(x, y, 0) = \sum_{k=0}^{[\frac{N}{2}]} c_k \mathrm{LG}_{k,\pm(N-2k)}(x, y) = \exp\left(-\frac{r^2}{w^2}\right) \sum_{k=0}^{[\frac{N}{2}]} c_k \left(\frac{x \pm iy}{w}\right)^{N-2k} L_k^{N-2k}\left(\frac{2r^2}{w^2}\right), \tag{1.48}$$

In Eq. (1.48), there are Laguerre polynomials in the sum, (r, φ) are the polar coordinates in the waist plane of the LG beam. The square brackets in Eq. (1.48) denote the integer part of a number. Below, we use a dimensionless radial variable r instead of the ratio r/w and limit our study by considering only positive-charge optical vortices, i.e., we choose the sign "+" in Eq. (1.48). The topological charge of the optical vortices superposition (1.48) can be determined by obtaining the argument (phase) of an asymptotic expression (1.48) when x and y tend to infinity. For this purpose, we apply an asymptotic expression for the Laguerre polynomials ($r >> 1$):

$$L_k^{N-2k}(2r^2) \approx \frac{(-2r^2)^k}{k!}. \tag{1.49}$$

Then, instead of Eq. (1.48), we write an asymptotic expression for the amplitude of beams (1.48):

$$E(r >> 1) = e^{-r^2+iN\varphi} r^N \sum_{k=0}^{[\frac{N}{2}]} c_k \frac{(-2)^k}{k!} \exp(-2ik\varphi). \tag{1.50}$$

It is difficult to obtain the TC of the superposition (1.50) for arbitrary real coefficients c_k. Therefore, we consider only several partial cases that allow for a closed-form representation of the sum in Eq. (1.50). We note that at certain coefficients values, the superposition of optical vortices (1.50) reduces to the geometrical sequence. The TC in the geometrical sequence of optical vortices was obtained in [64]. Therefore, we consider other examples of choosing the coefficients in Eq. (1.48). We use the well-known sum following from Newton's binomial theorem [63]:

$$\sum_{k=0}^{N} \frac{N!}{k!(N-k)!} \exp(-ik\varphi) = 2^N \left(\cos\frac{\varphi}{2}\right)^N \exp\left(-i\frac{N}{2}\varphi\right). \tag{1.51}$$

Then, choosing in Eq. (1.50) the following coefficients:

$$c_k = \frac{(N/2)!}{(-2)^k (N/2-k)!}, \tag{1.52}$$

we obtain instead of Eq. (1.50) the following expression:

$$\begin{aligned} E(r>>1) &= e^{-r^2+iN\varphi} r^N \sum_{k=0}^{[N/2]} \frac{(N/2)!}{k!(N/2-k)!} \exp(-2ik\varphi) \\ &= 2^{N/2} (\cos\varphi)^{N/2} e^{-r^2} r^N \exp\left(i\frac{N}{2}\varphi\right). \end{aligned} \tag{1.53}$$

As seen from Eq. (1.53), the argument of the asymptotic expression is equal to ($N\varphi$)/2, and the TC of the superposition (1.48) with the coefficients (1.52) is thus equal to $N/2$. As was shown in [36], the half-integer TC in the initial plane becomes integer and equal to N upon propagation of the beam (1.48). Similarly, other examples of the sums (1.51) can be considered, many of which were given in [63].

1.2.6 Numerical Simulation

A numerical simulation of the propagation of all the beams in this work was conducted by the Fresnel transform implemented as the convolution by using the fast Fourier transform. The obtained distributions we compared with the respective theoretical formulae for the complex amplitudes. Numerical and theoretical intensity distributions were visually indistinguishable, whereas the phase distributions were different only in the areas of negligible low intensity. The topological charge of all the considered beams was computed as a sum of the phase (argument of the complex amplitude) derivatives by the azimuthal angle, computed over a circle of a "large" radius with small light intensity, and divided by the complete angle of 2π. As seen in all the figures below, the beams are propagation-invariant, i.e., conserve their intensity structure and change only in scale.

Figure 1.7 illustrates the intensity and phase distributions of the HLG beams (1.31) at $(n, m) = (5, 3)$ for different values of the parameter θ: 0, $\pi/16$, $\pi/8$, $3\pi/16$, $\pi/4$ in the initial plane $z = 0$ and at the Rayleigh distance $z = z_0$. Shown in Fig. 1.8 are also the intensity and phase distributions of the HLG beams (1.31) for the same values of the parameter θ and for the same propagation distances z, but at $(n, m) = (2, 7)$. Both these figures confirm that, when θ changes from 0 till $\pi/4$, the HG beam with $(n + 1) \times (m + 1)$ light spots is converted into an LG beam with $\min(n, m) + 1$ light rings. Computation of the topological charge over the phase distributions from Fig. 1.7 yielded the following values: 0.000000 (Fig. 1.7f), 1.999466 (Fig. 1.7g), 1.999875 (Fig. 1.7h), 1.999973 (Fig. 1.7i), 1.999995 (Fig. 1.7j), 0.000001 (Fig. 1.7p), 1.994684

(Fig. 1.7q), 1.993242 (Fig. 1.7r), 1.992495 (Fig. 1.7s), 1.992265 (Fig. 1.7t). Thus, computation confirms that the TC is zero at $\theta = 0$ and the TC is equal to 2 ($n - m = 5 - 3 = 2$) for all other values of θ. Phase distributions in the initial plane, shown in Fig. 1.7, clearly indicate the TC values. On the dashed red circle in Fig. 1.7f, there are only phase jumps by $+\pi$ (when black color is changed to white, counter-clockwise) and by $-\pi$ (when white color is changed to black). In total, all these jumps yield zero. Thus, the TC of a beam from Fig. 1.7a, f is zero. As is clearly seen in Fig. 1.7g–i, there are four nonhorizontal rays, where the phase on the dashed circle is changed by $+2\pi$, and two horizontal rays, where the phase on the circle jumps by -2π. Thus, the TC of the beams from Fig. 1.7g–i equals $4 - 2 = 2$. Finally, in Fig. 1.7j, the dashed circle is intersected by two horizontal rays, where the phase is changed by 2π. Thus, the TC of the beam from Fig. 1.7j is equal to 2. The same way, the TC can be simply computed for all other figures in this work.

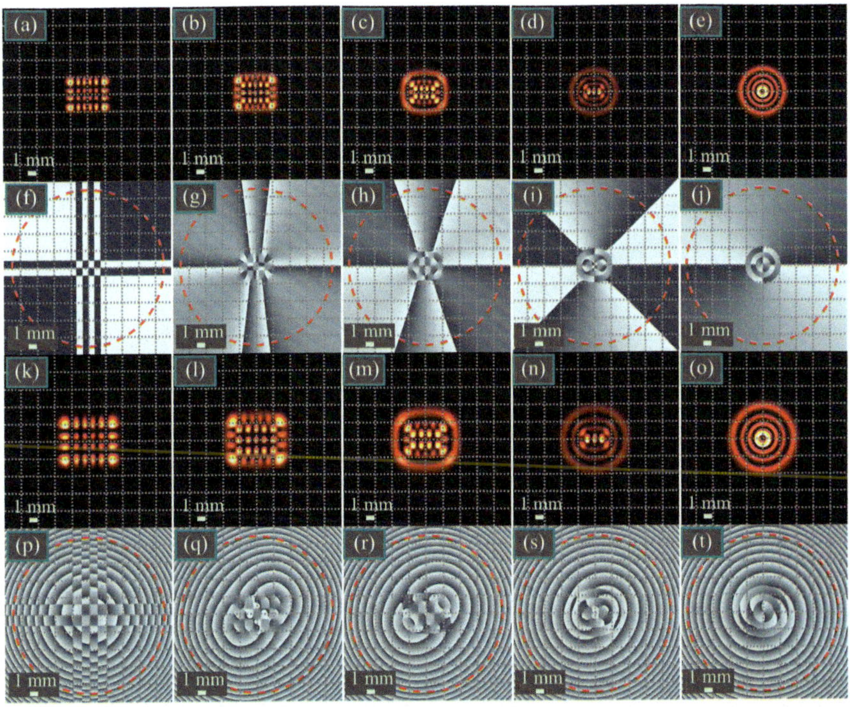

Fig. 1.7 Intensity (**a**–**e**, **k**–**o**) and phase (**f**–**j**, **p**–**t**) distributions of the Hermite–Laguerre-Gaussian beams (1.31) in the initial plane $z = 0$ (**a**–j) and at the Rayleigh distance $z = z_0$ (**k**–**t**) for different θ values. Computation parameters: wavelength $\lambda = 532$ nm, waist radius of the Gaussian beam $w_0 = 1$ mm, the beam order (orders of the Hermite polynomials) $(n, m) = (5, 3)$, values of the parameter θ: 0 (**a**, **f**, **k**, **p**), $\pi/16$ (**b**, **g**, **l**, **q**), $\pi/8$ (**c**, **h**, **m**, **r**), $3\pi/16$ (**d**, **i**, **n**, **s**), $\pi/4$ (**e**, **j**, **o**, **t**). Scale mark in each figure denotes 1 mm. The red dashed circles in the phase distributions are the circles over which the topological charge was computed. Black and white colors denote, respectively, the phase of 0 and π (f), or 0 and 2π (**g**–**j**, **p**–**t**)

For Fig. 1.8, the computed TC values are 0.000000 (Fig. 1.8f), −4.998308 (Fig. 1.8g), −4.999542 (Fig. 1.8h), −4.999854 (Fig. 1.8i), −4.999926 (Fig. 1.8j), −0.000001 (Fig. 1.8p), −4.986253 (Fig. 1.8q), −4.982896 (Fig. 1.8r), −4.981142 (Fig. 1.8s), and −4.980602 (Fig. 1.8t). This confirms that the TC is zero at $\theta = 0$ and the TC is −5 ($n - m = 2 - 7 = -5$) for all other θ values.

Combining the TC computation results found in Fig. 1.7 and Fig. 1.8, we obtain that the TC of the beam from Eq. (1.31) is equal to $n - m$ for an arbitrary value θ from the interval (0, π/2), which is consistent with the above theory.

Figure 1.9 depicts the intensity and phase distributions of an HLG beam (1.38) from Fig. 1.7c, h, m, r ($n = 5$, $m = 3$, $\theta = \pi/8$) for different values of the parameter α (1, 0.8, 0.6, 0.4, 0.2) in the initial plane $z = 0$ and at the Rayleigh distance $z = z_0$. According to Fig. 1.9, with a decreasing value of the parameter α, the HLG beam of the order (n, m) reduces into a Hermite beam of the order ($n + m$, 0). Computation of

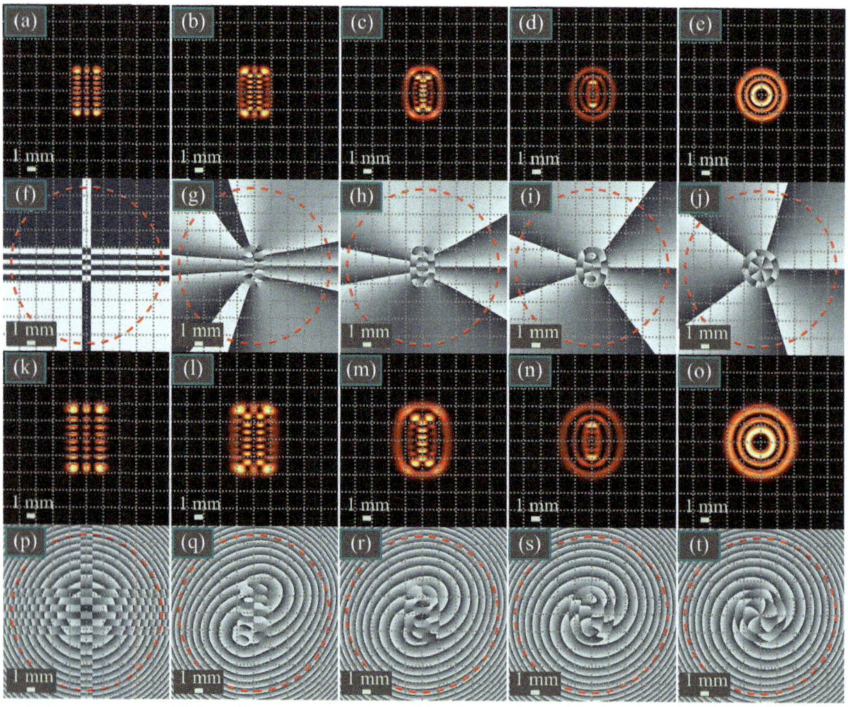

Fig. 1.8 Intensity (**a–e**, **k–o**) and phase (**f–j**, **p–t**) distributions of the Hermite–Laguerre-Gaussian beams (1.31) in the initial plane $z = 0$ (**a–j**) and at the Rayleigh distance $z = z_0$ (**k–t**) for different values θ. Computation parameters: wavelength $\lambda = 532$ nm, waist radius of the Gaussian beam w_0 = 1 mm, beam order (orders of the Hermite polynomials) (n, m) = (2, 7), values of the parameter θ: 0 (**a**, **f**, **k**, **p**), π/16 (**b**, **g**, **l**, **q**), π/8 (**c**, **h**, **m**, **r**), 3π/16 (**d**, **i**, **n**, **s**), π/4 (**e**, **j**, **o**, **t**). Scale mark in each figure denotes 1 mm. The red dashed circles in the phase distributions are the circles over which the topological charge was computed. Black and white colors denote, respectively, the phase of 0 and π (f), or 0 and 2π (**g–j**, **p–t**)

the topological charge over the phase distributions from Fig. 1.9 yielded the following values: 1.999871 (Fig. 1.9f), 1.999796 (Fig. 1.9g), 1.999654 (Fig. 1.9h), 1.999347 (Fig. 1.9i), 1.998661 (Fig. 1.9j), 1.993265 (Fig. 1.9p), 1.995472 (Fig. 1.9q), 1.998142 (Fig. 1.9r), 2.001121 (Fig. 1.9s), and 2.003653 (Fig. 1.9t). Thus, the computation confirms that the TC is equal to $n - m = 5 - 3 = 2$ at arbitrary values of α.

Shown in Fig. 1.10 are the intensity and phase distributions of a two-parametric vortex Hermite beam (1.41) at $(n, m) = (2, 7)$ for different values t and τ ((0.7, 0.7), (0.7, 0.4), (0.7, −0.3), (0.7, 0.2), (0.7, −0.1)) in the initial plane $z = 0$ and at the Rayleigh distance $z = z_0$. According to Eq. (1.41), when the magnitude of the parameter τ decreases, the beam should tend to the Hermite-Gaussian beam of the order $(n + m, 0)$. Figure 1.10 confirms this. Computation of the topological charge over the phase distributions from Fig. 1.10 yielded the following values: 8.999570 (Fig. 1.10f), 8.999303 (Fig. 1.10g), −8.998908 (Fig. 1.10h), 8.997925 (Fig. 1.10i), −8.994937 (Fig. 1.10j), 8.964787 (Fig. 1.10p), 8.966616 (Fig. 1.10q), −8.968667

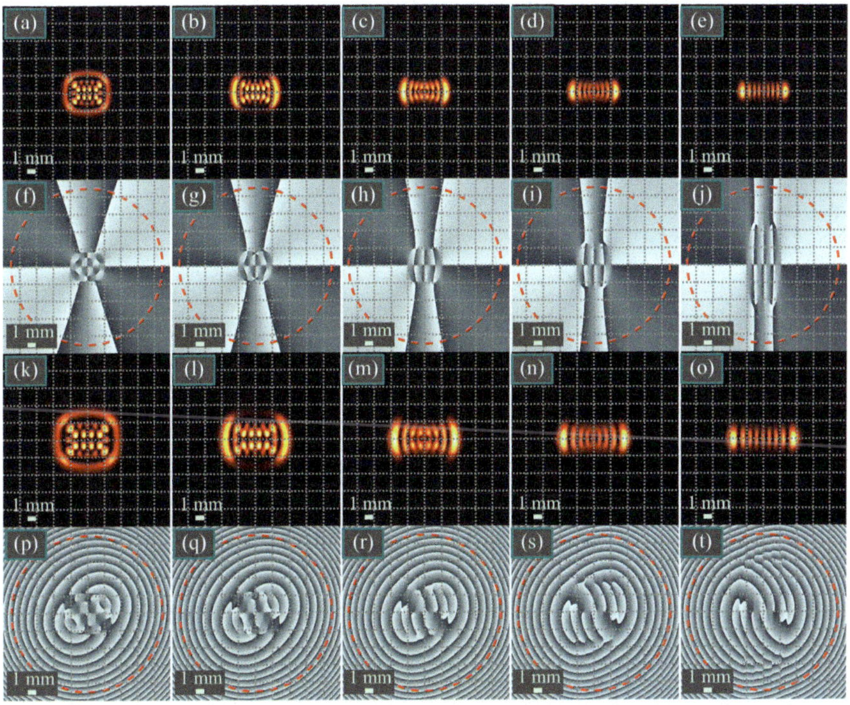

Fig. 1.9 Intensity (**a–e**, **k–o**) and phase (**f–j**, **p–t**) distributions of the Hermite–Laguerre-Gaussian beams (1.31) in the initial plane $z = 0$ (**a–j**) and at the Rayleigh distance $z = z_0$ (**k–t**) for different values of the parameter α. Computation parameters: wavelength $\lambda = 532$ nm, waist radius of the Gaussian beam $w_0 = 1$ mm, beam order (orders of the Hermite polynomials) $(n, m) = (5, 3)$, value of the parameter θ: $\theta = \pi/8$, values of the parameter α: 1 (**a**, **f**, **k**, **p**), 0.8 (**b**, **g**, **l**, **q**), 0.6 (**c**, **h**, **m**, **r**), 0.4 (**d**, **i**, **n**, **s**), 0.2 (**e**, **j**, **o**, **t**). Scale mark in each figure denotes 1 mm. The red dashed circles in the phase distributions are the circles over which the topological charge was

(Fig. 1.10r), 8.972072 (Fig. 1.10s), and −8.977350 (Fig. 1.10t). Thus, computation confirms that the TC is equal to $n + m = 2 + 7 = 9$ if $t\tau > 0$ and $-(n + m) = -(2 + 7) = -9$ if $t\tau < 0$.

Figure 1.11 illustrates the intensity and phase distributions of propagation-invariant superpositions of the Laguerre-Gaussian beams (1.48) for different orders N (4, 6, 8, 10, 12) in the initial plane $z = 0$ and at the Rayleigh distance $z = z_0$. Computation of the topological charge over the phase distributions from Fig. 1.11 yielded the following values: 3.999935 (Fig. 1.11f), 5.999776 (Fig. 1.11g), 7.999475 (Fig. 1.11h), 9.998990 (Fig. 1.11i), 11.998284 (Fig. 1.11j), 3.986310 (Fig. 1.11p), 5.979236 (Fig. 1.11q), 7.971959 (Fig. 1.11r), 9.964446 (Fig. 1.11s), and 11.956664 (Fig. 1.11t). Thus, computation confirms that the TC is equal to N.

In this section, the following results have been obtained. For the known family of propagation-invariant vortex Hermite–Laguerre–Gaussian laser beams, which are

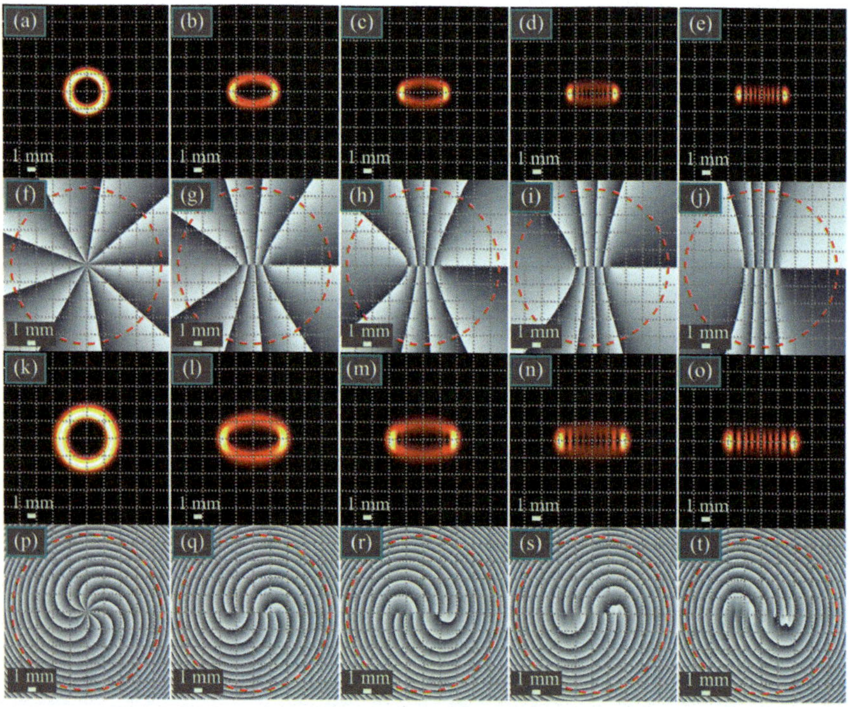

Fig. 1.10 Intensity (**a–e**, **k–o**) and phase (**f–j**, **p–t**) distributions of the two-parametric vortex Hermite beams (1.41) in the initial plane $z = 0$ (**a–j**) and at the Rayleigh distance $z = z_0$ (**k–t**) for different values of the parameters t and τ. Computation parameters: wavelength $\lambda = 532$ nm, waist radius of the Gaussian beam $w_0 = 1$ mm, beam order (orders of the Hermite polynomials) $(n, m) = (2, 7)$, values of the parameters (t, τ): (0.7, 0.7) (**a**, **f**, **k**, **p**), (0.7, 0.4) (**b**, **g**, **l**, **q**), (0.7, −0.3) (**c**, **h**, **m**, **r**), (0.7, 0.2) (**d**, **i**, **n**, **s**), (0.7, −0.1) (**e**, **j**, **o**, **t**). Scale mark in each figure denotes 1 mm. The red dashed circles in the phase distributions are the circles over which the topological charge was computed. Black and white colors denote, respectively, the phase of 0 and 2π (**f–j**, **p–t**)

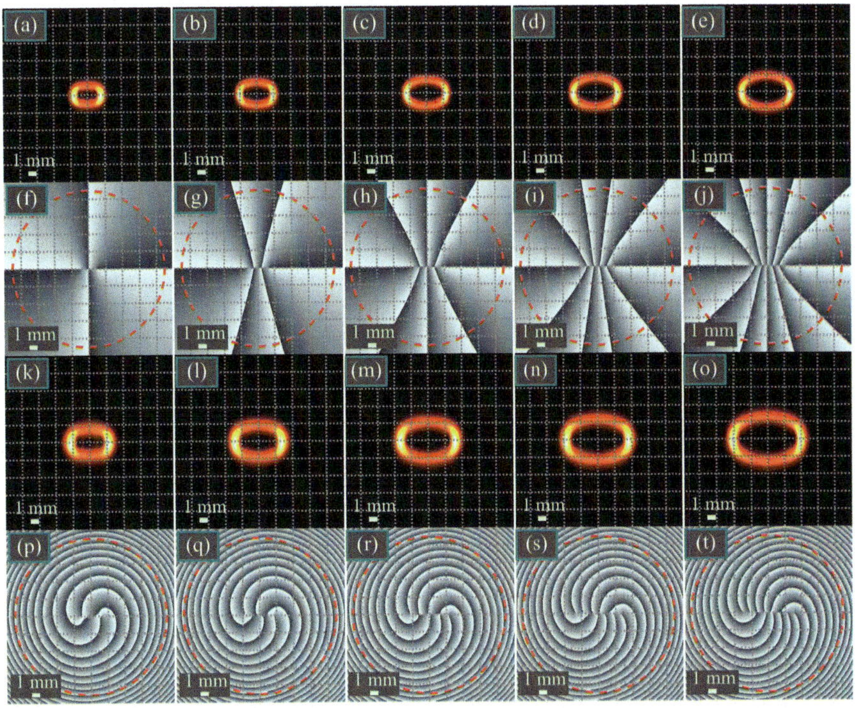

Fig. 1.11 Intensity (**a–e**, **k–o**) and phase (**f–j**, **p–t**) distributions of propagation-invariant superpositions of the Laguerre-Gaussian beams (1.48) in the initial plane $z = 0$ (**a–j**) and at the Rayleigh distance $z = z_0$ (**k–t**) for different beam orders N. Computation parameters: wavelength $\lambda = 532$ nm, waist radius of the Gaussian beam $w_0 = 1$ mm, beam order $N = 4$ (**a**, **f**, **k**, **p**), 6 (**b**, **g**, **l**, **q**), 8 (**c**, **h**, **m**, **r**), 10 (**d**, **i**, **n**, **s**), 12 (**e**, **j**, **o**, **t**). Scale mark in each figure denotes 1 mm. The red dashed circles in the phase distributions are the circles over which the topological charge was computed

finite superpositions of Hermite-Gaussian beams with a constant sum of indices of the Hermite polynomials $n + m$ and whose intensity shape depends on the parameter θ, we have demonstrated theoretically and numerically that, for an arbitrary value θ from the half-interval $(0, \pi/4]$, the topological charge of these beams is equal to $n - m$ [65]. If $\theta = 0$, the topological charge of the Hermite–Laguerre-Gaussian beam is zero. For another type of propagation-invariant beams, called two-parametric vortex Hermite beams, which are also finite superpositions of the Hermite beams with a constant sum of indices $n + m$ and with the partial amplitudes given by the binomial coefficients, we have shown theoretically and numerically that the topological charge is equal to the sum of indices $n + m$ if both parameters are of the same sign, and equal to $-(n + m)$ if the parameters are of different signs. We have also shown that if a propagation-invariant beam is a finite superposition of the Laguerre-Gaussian beams with the radial index p and with the azimuthal index $l > 0$, such that their combination $2p + l$ is constant, and if the amplitude multipliers are chosen as the binomial coefficients, then the topological charge of these beams is equal to $2p + l$.

This study has also revealed that the topological charge, which is one of the important characteristics of optical vortices along with the orbital angular momentum, is a quantity resistant to changing parameters of propagation-invariant beams. For instance, when the parameters θ and α of the Hermite–Laguerre-Gaussian beams are changed, the topological charge remains constant. The topological charge also remains unchanged upon the free-space propagation of beams. The obtained results will be useful for probing a weak-turbulence atmosphere by propagation-invariant vortex laser beams and for beam identification by measuring the beam topological charge, rather than the orbital angular momentum, since it remains constant at small beam distortions. The topological charge of vortex laser beams can be measured by using the Hartmann wavefront sensor. The simplest way to obtain the OAM of a laser beam is using a cylindrical lens.

References

1. H. Kogelnik, T. Li, Laser beams and resonators. Proc. IEEE **54**, 1312–1329 (1966). https://doi.org/10.1109/PROC.1966.5119
2. A.E. Siegman, Hermite-Gaussian functions of complex argument as optical beam eigenfunction. J. Opt. Soc. Am. **63**, 1093–1094 (1973)
3. R. Pratesi, L. Ronchi, Generalized Gaussian beams in free space. J. Opt. Soc. Am. **67**, 1274–1276 (1977)
4. E. Zauderer, Complex argument Hermite-Gaussian and Laguerre-Gaussian beams. J. Opt. Soc. Am. A **3**, 465–469 (1986)
5. A. Wünsche, Generalized Gaussian beam solutions of paraxial optics and their connection to a hidden symmetry. J. Opt. Soc. Am. A **6**, 1320–1329 (1989)
6. V.V. Kotlyar, A.A. Kovalev, Hermite-Gaussian modal laser beams with orbital angular momentum. J. Opt. Soc. Am. A **31**, 274–282 (2014)
7. Y. Wang, Y. Chen, Y. Zhang, H. Chen, S. Yu, Generalised Hermite-Gaussian beams and mode transformations. J. Opt. **18**, 055001 (2016)
8. E.G. Abramochkin, V.G. Volostnikov, Generalized Gaussian beams. J. Opt. A Pure Appl. Opt. **6**, S157–S161 (2004)
9. E.G. Abramochkin, V.G. Volostnikov, Beam transformations and nontransformed beams. Opt. Commun. **83**, 123–135 (1991)
10. V.V. Kotlyar, A.A. Kovalev, A.P. Porfirev, Vortex Hermite-Gaussian laser beams. Opt. Lett. **40**, 701–704 (2015)
11. H. Kogelnik, T. Li, Laser beams and resonators. Appl. Opt. **5**, 1550–1567 (1996)
12. Z. Mei, D. Zhao, J. Gu, H. Mao, Approximate analytical expressions of Laguerre-Gaussian beams passing through a paraxial optical system with an annular aperture. Optik **115**, 311–316 (2004)
13. J. Arlt, R. Kuhn, K. Dholakia, Spatial transformation of Laguerre-Gaussian laser modes. J. Mod. Opt. **48**, 783–787 (2001)
14. V. Jarutis, R. Paškauskas, A. Stabinis, Focusing of Laguerre-Gaussian beams by axicon. Opt. Commun. **184**, 105–112 (2000)
15. R. Simon, G.S. Agarwal, Wigner representation of Laguerre-Gaussian beams. Opt. Lett. **25**, 1313–1315 (2000)
16. H.C. Kim, Y.H. Lee, Hermite-Gaussian and Laguerre-Gaussian beams beyond the paraxial approximation. Opt. Commun. **169**, 9–16 (1999)
17. T. Hasegawa, T. Shimizu, Frequency-doubled Hermite-Gaussian beam and the mode conversion to the Laguerre-Gaussian beam. Opt. Commun. **160**, 103–108 (1999)

18. A.A. Kovalev, V.V. Kotlyar, A.P. Porfirev, Asymmetric Laguerre-Gaussian beams. Phys. Rev. A **93**, 063858 (2016)
19. Y.H. Hsieh, Y.H. Lai, M.X. Hsieh, K.F. Huang, Y.F. Chen, Generating high-power asymmetrical Laguerre-Gaussian modes and exploring topological charges distribution. Opt. Express **26**, 31738–31749 (2018)
20. M.G. Goran-Abad, M. Mahmoudi, Laguerre-Gaussian modes generated vector beam via nonlinear magneto-optical rotation. Sci. Rep. **11**, 5972 (2021)
21. Y. Yang, Y. Li, C. Wang, Generation and expansion of Laguerre-Gaussian beams. J. Opt. **51**, 910–926 (2022)
22. N. Matsumoto, T. Ando, T. Inoue, Y. Ohtake, N. Fukuchi, T. Hara, Generation of high-quality higher-order Laguerre-Gaussian beams using liquid-crystal-on-silicon spatial light modulators. J. Opt. Soc. Am. A **25**, 1642–1651 (2008)
23. L. Yu, Y. Zhang, J. Wang, Propagation of asymmetric Bessel mode in turbulent atmosphere. IEEE Photon. J. **14**, 7354106 (2022)
24. R.J. Watkins, K. Dai, G. White, W. Li, J.K. Miller, K.S. Morgan, E.G. Johnson, Experimental probing of turbulence using a continuous spectrum of asymmetric OAM beams. Opt. Express **28**, 924–935 (2020)
25. A.A. Kovalev, V.V. Kotlyar, A.P. Porfirev, Optical trapping and moving of microparticles by using asymmetrical Laguerre-Gaussian beams. Opt. Lett. **41**, 2426–2429 (2016)
26. M.A. Rykov, R.V. Skidanov, Modifying the laser beam intensity distribution for obtaining improved strength characteristics of an optical trap. Appl. Opt. **53**, 156–164 (2014)
27. A.P. Prudnikov, Y.A. Brychkov, OI. Marichev, *Integrals and Series. Special Functions* (Nauka Publisher, Moscow, 1983)
28. Y.A. Brychkov, *Special Functions. Derivatives, Integrals, Series, and other Formulae* (Fizmatlit Publisher, Moscow, 2006)
29. M.V. Berry, Optical vortices evolving from helicoidal integer and fractional phase steps. J. Opt. A Pure Appl. Opt. **6**, 259 (2004)
30. M. Abramowitz, I.A. Stegun, *Handbook of Mathematical Functions: With Formulas, Graphs, and Mathematical Tables* (National Bureau of Standards, Gaithersburg, 1965)
31. E.G. Abramochkin, V.G. Volostnikov, Spiral light beams,". Phys. Usp. **47**, 1177–1203 (2004)
32. E.G. Abramochkin, V.V. Kotlyar, A.A. Kovalev, S.S. Stafeev, Generalized asymmetric Hermite-Gaussian and Laguerre-Gaussian beams. Photonics **10**, 606 (2023)
33. E. Abramochkin, E. Razueva, V. Volostnikov, General astigmatic transform of Hermite–Laguerre-Gaussian beams. J. Opt. Soc. Am. A **27**, 2506–2513 (2010)
34. D. Deng, Q. Guo, W. Hu, Hermite–Laguerre-Gaussian beams in strongly nonlocal nonlinear media. J. Phys. B At. Mol. Opt. Phys. **41**, 225402 (2008)
35. Y.Q. Xu, G.Q. Zhou, X.G. Wang, Nonparaxial propagation of Hermite–Laguerre-Gaussian beams in uniaxial crystal orthogonal to the optical axis. Chin. Phys. B **22**, 064101 (2013)
36. K. Duan, B. Lü, Propagation of Hermite–Laguerre-Gaussian beams through a paraxial optical ABCD system with rectangular hard-edged aperture. Opt. Commun. **250**, 1–9 (2005)
37. D. Deng, Q. Guo, Elegant Hermite–Laguerre-Gaussian beams. Opt. Lett. **33**, 1225–1227 (2008)
38. E. Abramochkin, T. Alieva, Closed-form expression for mutual intensity evolution of Hermite–Laguerre-Gaussian Schell-model beams. Opt. Lett. **42**, 4032–4035 (2017)
39. J. Zhang, X. Yu, Y. Chen, M. Huang, X. Dai, D. Liu, Generation of Hermite–Laguerre-Gaussian beams based on space-variant Pancharatnam Berry phase. Proc. SPIE **10964**, 109645R (2018)
40. E. Abramochkin, V. Volostnikov, Spiral-type beams: optical and quantum aspects. Opt. Commun. **125**, 302–323 (1996)
41. E.G. Abramochkin, V.G. Volostnikov, *Modern optics of Gaussian Beams* (Fizmatlit, Moscow, 2010)
42. V.V. Kotlyar, A.A. Kovalev, Orbital angular momentum of paraxial propagation-invariant laser beams. J. Opt. Soc. Am. A **39**, 1061–1065 (2022)
43. A. Zannotti, C. Denz, M.A. Alonso, M.R. Dennis, Shaping caustics into propagation-invariant light. Nat. Commun. **11**, 3597 (2020)

44. M. Soskind, R. Soskind, Y. Soskind, Shaping propagation invariant laser beams. Opt. Eng. **54**, 111309 (2015)
45. A. Hansen, J.T. Schultz, N.P. Bigelow, Singular atom optics with spinor Bose-Einstein condensates. Optica **3**, 355–361 (2016)
46. J. Wang, J.Y. Yang, I.M. Fazal, N. Ahmed, Y. Yan, H. Huang, Y. Ren, Y. Yue, S. Dolinar, M. Tur, A.E. Willner, Terabit free-space data transmission employing orbital angular momentum multiplexing. Nat. Photon. **6**, 488–496 (2012)
47. J.P. Torres, Multiplexing twisted light. Nat. Photon. **6**, 420–422 (2012)
48. Z. Xie, T. Lei, F. Li, H. Qiu, Z. Zhang, H. Wang, C. Min, L. Du, Z. Li, X. Yuan, Ultra-broadband on-chip twisted light emitter for optical communications. Light Sci. Appl. **7**, 18001 (2018)
49. A. Sit, F. Bouchard, R. Fickler, J. Gagnon-Bischoff, H. Larocque, K. Heshami, D. Elser, C. Peuntinger, K. Günthner, B. Heim, C. Marquardt, G. Leuchs, R.W. Boyd, E. Karimi, High-dimensional intracity quantum cryptography with structured photons. Optica **4**, 1006–1010 (2017)
50. A. Sit, R. Fickler, F. Alsaiari, F. Bouchard, H. Larocque, P. Gregg, L. Yan, R.W. Boyd, S. [illegible], E. Karimi, Quantum cryptography with structured photons through a vortex fiber. Opt. Lett. **43**, 4108–4111 (2018)
51. M. Erhard, R. Fickler, M. Krenn, A. Zeilinger, Twisted photons: new quantum perspectives in high dimensions. Light Sci. Appl. **7**, 17146 (2018)
52. A. Mathis, F. Courvoisier, L. Froehly, L. Furfaro, M. Jacquot, P.A. Lacourt, J.M. Dudley, Micro-machining along a curve: Femtosecond laser micromachining of curved profiles in diamond and silicon using accelerating beams. Appl. Phys. Lett. **101**, 071110 (2012)
53. F. Courvoisier, R. Stoian, A. Couairon, Ultrafast laser micro-and nano-processing with nondiffracting and curved beams. Opt. Laser Technol. **80**, 125–137 (2016)
54. S.W. Hell, J. Wichmann, Breaking the diffraction resolution limit by stimulated emission: stimulated-emission-depletion fluorescence microscopy. Opt. Lett. **19**, 780–782 (1994)
55. K.I. Willig, B. Harke, R. Medda, S.W. Hell, STED microscopy with continuous wave beams. Nat. Methods **4**, 915–918 (2007)
56. F.Q. Fahrbach, P. Simon, A. Rohrbach, Microscopy with self-reconstructing beams. Nat. Photon. **4**, 780–785 (2010)
57. T. Vettenburg, H.I. Dalgarno, J. Nylk, C. Coll-Lladó, D.E. Ferrier, T. Čižmár, F.J. Gunn-Moore, K. Dholakia, Light-sheet microscopy using an Airy beam. Nat. Methods **11**, 541–544 (2014)
58. K. Dholakia, T. Čižmár, Shaping the future of manipulation. Nat. Photon. **5**, 335–342 (2011)
59. M. Woerdemann, C. Alpmann, M. Esseling, C. Denz, Advanced optical trapping by complex beam shaping. Laser Photon. Rev. **7**, 839–854 (2013)
60. A. Volyar, E. Abramochkin, Y. Egorov, M. Bretsko, Y. Akimova, Fine structure of perturbed Laguerre-Gaussian beams: Hermite-Gaussian mode spectra and topological charge. Appl. Opt. **59**, 7680–7687 (2020)
61. A. Volyar, E. Abramochkin, Y. Akimova, M. Bretsko, Super bursts of the orbital angular momentum in astigmatic-invariant structured LG beams. Opt. Lett. **47**, 5537–5540 (2022)
62. A. Volyar, E. Abramochkin, Y. Akimova, M. Bretsko, Control of the orbital angular momentum via radial numbers of structured Laguerre-Gaussian beams. Opt. Lett. **47**, 2402–2405 (2022)
63. A.P. Prudnikov, Y.A. Brychkov, O.I. Marichev, *Integrals and Series. Volume 1: Elementary Functions* (Gordon and Breach, New York, 1986)
64. V. Kotlyar, A. Kovalev, E. Kozlova, A. Savelyeva, S. Stafeev, Geometric progression of optical vortices. Photonics **9**, 407 (2022)
65. V.V. Kotlyar, A.A. Kovalev, E.G. Abramochkin, Topological charge of propagation-invariant laser beams. Photonics **10**, 915 (2023)

Chapter 2
Ince-Gaussian Beams

2.1 Structurally Invariant Higher-Order Ince-Gaussian Beams and Their Expansions into the Hermite-Gaussian or Laguerre-Gaussian Beams

When using light for steering micromachine components and generating multiple optical traps for microscopic particles, laser beam modes offer a convenient tool as they conserve their structure both upon free-space propagation and at the focus of a spherical lens. Another benefit is that mode symmetry can be adjusted by varying certain parameters and passing from symmetry in the Cartesian coordinates to circular symmetry via elliptical symmetry. For this purpose, Ince-Gaussian (IG) modes can be utilized, as they are transformed into either Hermite-Gaussian (HG) beams or Laguerre-Gaussian (LG) beams when the ellipticity parameter is changed.

A solution of the Whittaker equation in the form of IG functions was obtained for the first time by Arscott [1, 2]. In the book by Miller [3], this solution was obtained in separated variables in elliptical coordinates. In optics, IG beams were considered in [4]. In [5], expressions for the amplitudes of the IG beams have been obtained for $p = 0, 1, 2$ when the IG beams have no dependence of the ellipticity parameter. In [6], elegant IG beams were investigated. These works have given an impetus for a wide use of IG beams in optics, along with other well known laser modes (LG and HG). In [7], IG beams were generated with a digital hologram. In [8], vector (classically entangled) IG beams were generated as superposition of a right-handed circularly polarized even IG beam with a left-handed circularly polarized odd IG beam. The IG beams with quantum entanglement were generated in [9]. Besides, nonlinear transformation of the IG beams by spontaneous parametric down-conversion was investigated in [10, 11]. In [12], IG beams were generated also parametrically, but without conversion. In [13], elegant IG beams were studied in a parabolic medium. Propagation of the IG beams in uniaxial crystals was studied in [14]. The IG beams

V. V. Kotlyar et al., *Helical Laser Beams*, https://doi.org/10.1007/978-3-031-92217-6_2

are used for underwater data transmission [15]. Propagation of the IG beams in a turbulent atmosphere was considered in [16].

As seen from the above brief review of works dealing with the IG beams, they have been actively studied in optics. However, no analytical representation of these beams via the LG or HG modes with an explicit dependence on the ellipticity parameter has yet been proposed. In this section, in an effort to fill in this gap, we derive a number of particular analytical formulae for even and odd IG modes with the index $p = 3, 4, 5, 6$, which are expressed via the LG and HG modes and contain explicit dependence on the ellipticity parameter ε. This explicit dependence of the expansion coefficients on the ellipticity parameter allows controlling the intensity pattern of the IG beams by continuously varying this parameter. In this work, for the first time in optics we consider a situation when the ellipticity parameter ε can be not only positive, but negative as well. We analyze in which way the IG beam changes with changing sign of this parameter. Simultaneous representation of the IG modes via several LG modes and several HG modes proposed herein, suggests that when the ellipticity parameter ε tends to zero, any IG mode is converted into a certain LG mode (more exactly, to its real or imaginary part), and when the ellipticity parameter ε tends to infinity, the IG mode coincides with a certain HG mode.

2.1.1 *Solution of the Paraxial Equation in Elliptic Coordinates*

The paraxial Helmholtz equation is given by [3]:

$$\left(\frac{\partial^2}{\partial x^2} + \frac{\partial^2}{\partial y^2} + 2ik\frac{\partial}{\partial z}\right)E(\mathbf{r}, z) = 0, \tag{2.1}$$

where (x, y, z) are the Cartesian coordinates, $\mathbf{r} = (x, y) = (r\cos\varphi, r\sin\varphi)$ is a 2D vector, z is the coordinate along the optical axis, k is the wavenumber of light, $E(\mathbf{r}, z)$ is the complex amplitude of a monochromatic light field. Below, we use dimensionless coordinates: $x := x/w_0$, $y := y/w_0$, $z := 2z/(kw_0^2)$, with w_0 being the waist radius of the Gaussian beam. Then, in dimensionless variables, Eq. (2.1) reads as

$$\left(\frac{\partial^2}{\partial x^2} + \frac{\partial^2}{\partial y^2} + 4i\frac{\partial}{\partial z}\right)E(\mathbf{r}, z) = 0. \tag{2.2}$$

Its simplest solution with a finite energy is a conventional Gaussian beam:

$$G(\mathbf{r}, z) = \frac{1}{1+iz}\exp\left(-\frac{r^2}{1+iz}\right). \tag{2.3}$$

It is known that separation of variables in the Eq. (2.2) in the Cartesian and polar coordinates allows obtaining two solution families—HG and LG beams, respectively [17]:

$$\mathcal{H}_{n,m}(\boldsymbol{r}, z) = G(\boldsymbol{r}, z)\left(\frac{1-iz}{1+iz}\right)^{\frac{n+m}{2}} H_n\left(\frac{x\sqrt{2}}{\sqrt{1+z^2}}\right)H_m\left(\frac{y\sqrt{2}}{\sqrt{1+z^2}}\right),$$
$$\mathcal{L}_{n,\pm m}(\boldsymbol{r}, z) = G(\boldsymbol{r}, z)\left(\frac{1-iz}{1+iz}\right)^{n+\frac{m}{2}} L_n^m\left(\frac{2r^2}{1+z^2}\right)r^m e^{\pm im\varphi}, \tag{2.4}$$

with $n, m = 0, 1, 2, \ldots$.

Both families are examples of structurally stable light fields, i.e. at any propagation distance z, the transverse intensity pattern of such fields coincides—up to scale—with that in the initial plane $z = 0$:

$$\mathcal{H}_{n,m}(\mathbf{r}, z) = \frac{1}{\sqrt{1+z^2}}\mathcal{H}_{n,m}\left(\frac{\mathbf{r}}{\sqrt{1+z^2}}, 0\right)\exp\left(\frac{izr^2}{1+z^2} - i(n+m+1)\arctan z\right),$$
$$\mathcal{L}_{n,\pm m}(\mathbf{r}, z) = \frac{1}{\sqrt{1+z^2}}\mathcal{L}_{n,\pm m}\left(\frac{\mathbf{r}}{\sqrt{1+z^2}}, 0\right)\exp\left(\frac{izr^2}{1+z^2} - i(2n+m+1)\arctan z\right). \tag{2.5}$$

Thus, it will suffice to study both families at $z = 0$, omitting the last argument for brevity: $\mathcal{H}_{n,m}(\mathbf{r}, 0) = \mathcal{H}_{n,m}(\mathbf{r})$ and $\mathcal{L}_{n,\pm m}(\mathbf{r}, 0) = \mathcal{L}_{n,\pm m}(\mathbf{r})$.

We also introduce normalized variants of each solutions family:

$$\mathbf{HG}_{n,m}(\mathbf{r}) = \frac{\mathcal{H}_{n,m}(\mathbf{r})}{\left\|\mathcal{H}_{n,m}(\mathbf{r})\right\|}, \quad \left\|\mathcal{H}_{n,m}(\mathbf{r})\right\| = \sqrt{\frac{\pi}{2}\cdot 2^{n+m}n!m!},$$
$$\mathbf{LG}_{n,\pm m}(r) = \frac{\mathcal{L}_{n,\pm m}(\mathbf{r})}{\left\|\mathcal{L}_{n,\pm m}(\mathbf{r})\right\|}, \quad \left\|\mathcal{L}_{n,\pm m}(\mathbf{r})\right\| = \sqrt{\frac{\pi}{2}\cdot\frac{(n+m)!}{2^m n!}}. \tag{2.6}$$

When solving the Eq. (2.2), the variables are separated in the elliptic coordinates (ξ, η) [3], then a family of IG beams is obtained, which are also structurally stable:

$$\mathcal{I}(\mathbf{r}, z) = G(\mathbf{r}, z)\left(\frac{1-iz}{1+iz}\right)^{p/2} P(\xi)Q(\eta). \tag{2.7}$$

Here,

$$\frac{x}{\sigma\sqrt{1+z^2}} = \cosh\xi\cos\eta, \quad \frac{y}{\sigma\sqrt{1+z^2}} = \sinh\xi\sin\eta, \tag{2.8}$$

and the parameter σ is supposed to be arbitrary yet. Substitution of Eq. (2.7) into Eq. (2.2) yields equations for the functions $P(\xi)$ and $Q(\eta)$:

$$\left(\frac{\partial^2}{\partial \xi^2} - 2\sigma^2 \sinh 2\xi \frac{\partial}{\partial \xi} - (\lambda - 2p\sigma^2 \cosh 2\xi)\right)P(\xi) = 0,$$
$$\left(\frac{\partial^2}{\partial \eta^2} + 2\sigma^2 \sin 2\eta \frac{\partial}{\partial \eta} + (\lambda - 2p\sigma^2 \cos 2\eta)\right)Q(\eta) = 0. \tag{2.9}$$

Both Eq. (2.9) reduce to the Ince equation, with its canonical form being given by [1–3]:

$$\left(\frac{\partial^2}{\partial t^2} + \varepsilon \sin 2t \frac{\partial}{\partial t} + (\lambda - p\varepsilon \cos 2t)\right)N(t) = 0, \tag{2.10}$$

where p is an integer nonnegative number, ε is a positive ellipticity parameter, mentioned in the Introduction, and $\lambda = \lambda(p, \varepsilon)$ is some constant. In works [1–3] it is supposed that $\varepsilon > 0$. This is also seen from comparison of Eqs. (2.9) and (2.10) since $\varepsilon = 2\sigma^2 > 0$. Below in the current work, we show that the parameter ε in the solution (2.10) can be selected of any sign. In Eq. (2.9), on separating the variables ξ and η in Eq. (2.2) there appears a constant. For Eq. (2.10), the constant is a real number, introduced for the existence of solution $N(t)$ in the form of a trigonometric polynomial of degree p (called the Ince polynomial). Once the form of the solution is restricted in this way, we find that λ should be a root of a certain polynomial of degree $(p + 1)$, whose coefficients depend on ε (i.e. root of a characteristic equation). Thus, in total, $(p + 1)$ roots appear: $\lambda_0, \lambda_1, \ldots, \lambda_p$, and they are all real-valued. In addition, if these roots are sorted ascending at $\varepsilon = 0$, they are not mixing with increasing ε, i.e. their ascending order does not change (for any positive ε, the characteristic equation has no multiple roots). Thus, by choosing a root $\lambda_q(p, \varepsilon)$ and by constructing a trigonometric Ince polynomial $N(t) = N_{p,q}(t, \varepsilon)$, the solution (2.7) can be written via the solutions of the Eq. (2.10) as follows:

$$\mathcal{I}_{p,q}(\mathbf{r}, z) = G(\mathbf{r}, z)\left(\frac{1 - iz}{1 + iz}\right)^{p/2} N_{p,q}(i\xi, \varepsilon)N_{p,q}(\eta, \varepsilon), \tag{2.11}$$

with $\varepsilon = 2\sigma^2$. It can be shown that the characteristic equation is factorized and splits into two equations. If p is even then there are $p/2$ odd Ince polynomials (they are written via the sines of multifold angles and usually indexed as $q = 2, 4, \ldots, p$) and $p/2 + 1$ even Ince polynomials (they are written via the cosines of multifold angles and indexed as $q = 0, 2, 4, \ldots, p$). If p is odd, there are equally $(p + 1)/2$ even and odd Ince polynomials (indexed as $q = 1, 3, \ldots, p$). Thus, even and odd IG modes (2.11) are usually written as:

$$\mathbf{IG}^e_{p,q}(\mathbf{r}, z) = G(\mathbf{r}, z)\left(\frac{1 - iz}{1 + iz}\right)^{p/2} C_p^{(q)}(i\xi, \varepsilon)C_p^{(q)}(\eta, \varepsilon),$$
$$\mathbf{IG}^o_{p,q}(\mathbf{r}, z) = G(\mathbf{r}, z)\left(\frac{1 - iz}{1 + iz}\right)^{p/2} S_p^{(q)}(i\xi, \varepsilon)S_p^{(q)}(\eta, \varepsilon), \tag{2.12}$$

where $C_p^{(q)}(t, \varepsilon)$ and $S_p^{(q)}(t, \varepsilon)$ are even and odd Ince polynomials, respectively, and indices p and q are nonnegative integer numbers ($q \le p$).

Unfortunately, a tradition to write down the even and odd IG modes differently and study them separately leads to unjustified complications in the designations: the index q is chosen of the same parity as p, i.e. $q = \{p, p-2, p-4, \dots\}$. If p is odd then this list stops at $q = 1$, but if p is even then it stops at $q = 0$ for modes IG^e and at $q = 2$ for modes IG^o. This separation is probably related with the initial form of the Ince polynomials $N(t)$ expressed via $\sin kt$ or $coskt$. If, instead of the sines and cosines, exponentials e^{ikt} were used and if the IG modes were not separated into even and odd, then, for any fixed p, the index q would range from 0 till p, respective numbers $\lambda_q(p, \varepsilon)$ would be sorted ascending, and this would allow investigating the IG modes in a unified way, similarly to the Hermite-Gaussian modes, for which, despite the Hermite polynomials being both even and odd, the family of the HG modes is not split into the even and odd subfamilies. However, at this point, we will adhere to the commonly used designations.

Since the IG modes are structurally stable, below we write them at $z = 0$, omitting the argument z for brevity, but specifying explicitly the dependence on the parameter ε: $\mathbf{IG}_{p,q}^{(e,o)}(\mathbf{r}, z) \to \mathbf{IG}_{p,q}^{(e,o)}(\mathbf{r}) \to \mathbf{IG}_{p,q}^{(e,o)}(\mathbf{r}, \varepsilon)$. Such a notation is convenient when studying limiting cases $\varepsilon = 0$ and $\varepsilon = \infty$, when the IG modes reduce to the LG and HG modes, respectively. In addition, similarly to Eq. (2.6), we use the IG modes with and without normalization:

$$\mathbf{IG}_{p,q}^{(e,o)}(\mathbf{r}, \varepsilon) = \frac{\mathcal{I}_{p,q}^{(e,o)}(\mathbf{r}, \varepsilon)}{\left\| \mathcal{I}_{p,q}^{(e,o)}(\mathbf{r}, \varepsilon) \right\|}, \tag{2.13}$$

preferring the non-normalized variant in cumbersome cases, $\mathcal{I}_{p,q}^{(e,o)}(\mathbf{r}, \varepsilon)$, since for the IG modes, it is generally much simpler than the normalized one.

2.1.2 Expansions of the IG Modes by the HG and LG Modes at Small Values of P

The simplest case is when $p = 0$. Then, $q = 0$ and the even IG mode reduces to the conventional Gaussian beam:

$$\mathbf{IG}_{0,0}^{e}(\mathbf{r}, \varepsilon) = \mathbf{HG}_{0,0}(\mathbf{r}) = \mathbf{LG}_{0,0}(\mathbf{r}). \tag{2.14}$$

The following three modes are also independent of the parameter ε and were given in [4, 5]:

$$\mathbf{IG}_{1,1}^{e}(\mathbf{r}, \varepsilon) = \mathbf{HG}_{1,0}(\mathbf{r}) = \sqrt{2}\mathrm{Re}\mathbf{LG}_{0,1}(\mathbf{r}),$$
$$\mathbf{IG}_{1,1}^{o}(\mathbf{r}, \varepsilon) = \mathbf{HG}_{0,1}(\mathbf{r}) = \sqrt{2}\mathrm{Im}\mathbf{LG}_{0,1}(\mathbf{r}),$$

$$\mathbf{IG}^{o}_{2,2}(\mathbf{r}, \varepsilon) = \mathbf{HG}_{1,1}(\mathbf{r}) = \sqrt{2}\mathrm{Im}\mathbf{LG}_{0,2}(\mathbf{r}). \tag{2.15}$$

Cases with the Characteristic Polynomial of Degree 2

Below, without derivation, we give explicit analytical expressions for those IG modes, for which the characteristic polynomial reduces to a quadratic polynomial and the respective roots $\lambda_q(p, \varepsilon)$ can be expressed in square radicals.

It is worth noting that if the initial definition $\varepsilon = 2\sigma^2$ is dropped. the parameter ε (ellipticity parameter) can be considered to be a negative number (as noticed by Edward Ince [18]). Therefore, in addition to the limiting cases of the IG modes with $\varepsilon = 0$ and $\varepsilon = \infty$, we also consider a case when $\varepsilon = -\infty$.

In Table 2.1 (and in Tables 2.3 and 2.7 below), the following designations are used: $\mathbf{HG}_{n,m}(\mathbf{r})$ and $\mathbf{LG}_{n,l}(\mathbf{r})$ are the normalized Hermite-Gaussian and Laguerre-Gaussian modes (2.6), and $\mathbf{IG}^{(e,o)}_{p,q}(\mathbf{r}, \varepsilon)$ are the normalized Ince-Gaussian modes (2.13).

The coefficients a, b, c, d here are positive, monotonically increasing functions of ε:

$$\begin{aligned} a &= \varepsilon + \sqrt{\varepsilon^2 + 1}, \quad b = \varepsilon - 1 + \sqrt{(\varepsilon - 1)^2 + 3}, \\ c &= \varepsilon + 1 + \sqrt{(\varepsilon + 1)^2 + 3}, \quad d = \left(\varepsilon + \sqrt{\varepsilon^2 + 9}\right) \Big/ 3. \end{aligned} \tag{2.16}$$

We should note a similarity of the IG modes with equal indices p and parity flags (e, o). For instance, for obtaining the mode $\mathbf{IG}^{o}_{4,4}(\mathbf{r}, \varepsilon)$ from the mode $\mathbf{IG}^{o}_{4,2}(\mathbf{r}, \varepsilon)$, it is sufficient to swap the superposition coefficients and change the sign of one of them. It turns out that a similar property is valid in a more general case. Namely, if for fixed values p and (e, o) the whole set of IG modes is known, excepting one mode, then this unknown mode can be obtained by manipulating the expansion coefficients of the modes from the set. When the characteristic polynomial is quadratic, there are only two modes and the "manipulation" is just a swapping of the coefficients with a sign change. For more complicated cases, manipulations are also more complicated. However, they are incomparably simpler than obtaining the dependence of the superposition coefficients on the ellipticity parameter.

Cases with the Characteristic Polynomial of Degree 3

For the mode $\mathbf{IG}^{e}_{4,2m}(\mathbf{r}, \varepsilon)$ with $m = 0, 1, 2$, we get the following characteristic equation:

$$\chi^{e}_{4}(\lambda, \varepsilon) = \lambda^3 - 20\lambda^2 + 16(4 - \varepsilon^2)\lambda + 192\varepsilon^2 = 0, \tag{2.17}$$

whose roots $\lambda^{e}_{m}(\varepsilon)$ can be found by using the Cardano formula. In the ascending order they read as

$$\lambda^{e}_{0} = \tfrac{20}{3} - 2c_4, \quad \lambda^{e}_{1} = \tfrac{20}{3} + c_4 - s_4\sqrt{3}, \quad \lambda^{e}_{2} = \tfrac{20}{3} + c_4 + s_4\sqrt{3} \tag{2.18}$$

Table 2.1 Normalized IG beams expressed as superpositions of two HG beams or two LG beams, and their limiting cases

	$\varepsilon \to -\infty$	$\varepsilon = 0$	$\varepsilon \to +\infty$
$\mathbf{IG}^e_{2,0}(\mathbf{r},\varepsilon) = -\frac{\mathbf{HG}_{2,0}(\mathbf{r}) + a\cdot\mathbf{HG}_{0,2}(\mathbf{r})}{\sqrt{1+a^2}} = \frac{(a-1)\mathrm{Re}\mathbf{LG}_{0,2}(\mathbf{r}) + (a+1)\mathbf{LG}_{1,0}(\mathbf{r})}{\sqrt{2}\cdot\sqrt{1+a^2}}$	$-\mathbf{HG}_{2,0}(\mathbf{r})$	$\mathbf{LG}_{1,0}(\mathbf{r})$	$-\mathbf{HG}_{0,2}(\mathbf{r})$
$\mathbf{IG}^e_{2,2}(\mathbf{r},\varepsilon) = +\frac{a\cdot\mathbf{HG}_{2,0}(\mathbf{r}) - \mathbf{HG}_{0,2}(\mathbf{r})}{\sqrt{1+a^2}} = \frac{(a+1)\sqrt{2}\mathrm{Re}\mathbf{LG}_{0,2}(\mathbf{r}) - (a-1)\mathbf{LG}_{1,0}(\mathbf{r})}{\sqrt{2}\cdot\sqrt{1+a^2}}$	$-\mathbf{HG}_{0,2}(\mathbf{r})$	$\sqrt{2}\mathrm{Re}\mathbf{LG}_{0,2}(\mathbf{r})$	$+\mathbf{HG}_{2,0}(\mathbf{r})$
$\mathbf{IG}^e_{3,1}(\mathbf{r},\varepsilon) = -\frac{\sqrt{3}\cdot\mathbf{HG}_{3,0}(\mathbf{r}) + b\cdot\mathbf{HG}_{1,2}(\mathbf{r})}{\sqrt{3+b^2}} = \frac{\sqrt{3}(b-1)\mathrm{Re}\mathbf{LG}_{0,3}(\mathbf{r}) + (b+3)\mathrm{Re}\mathbf{LG}_{1,1}(\mathbf{r})}{\sqrt{2}\cdot\sqrt{3+b^2}}$	$-\mathbf{HG}_{3,0}(\mathbf{r})$	$\sqrt{2}\mathrm{Re}\mathbf{LG}_{1,1}(\mathbf{r})$	$-\mathbf{HG}_{1,2}(\mathbf{r})$
$\mathbf{IG}^e_{3,3}(\mathbf{r},\varepsilon) = +\frac{b\cdot\mathbf{HG}_{3,0}(\mathbf{r}) - \sqrt{3}\cdot\mathbf{HG}_{1,2}(\mathbf{r})}{\sqrt{3+b^2}} = \frac{(b+3)\mathrm{Re}\mathbf{LG}_{0,3}(\mathbf{r}) - \sqrt{3}(b-1)\mathrm{Re}\mathbf{LG}_{1,1}(\mathbf{r})}{\sqrt{2}\cdot\sqrt{3+b^2}}$	$-\mathbf{HG}_{1,2}(\mathbf{r})$	$\sqrt{2}\mathrm{Re}\mathbf{LG}_{0,3}(\mathbf{r})$	$+\mathbf{HG}_{3,0}(\mathbf{r})$
$\mathbf{IG}^o_{3,1}(\mathbf{r},\varepsilon) = -\frac{\sqrt{3}\cdot\mathbf{HG}_{2,1}(\mathbf{r}) + c\cdot\mathbf{HG}_{0,3}(\mathbf{r})}{\sqrt{3+c^2}} = \frac{(c-3)\mathrm{Im}\mathbf{LG}_{0,3}(\mathbf{r}) + \sqrt{3}(c+1)\mathrm{Im}\mathbf{LG}_{1,1}(\mathbf{r})}{\sqrt{2}\cdot\sqrt{3+c^2}}$	$-\mathbf{HG}_{2,1}(\mathbf{r})$	$\sqrt{2}Im\mathbf{LG}_{1,1}(\mathbf{r})$	$-\mathbf{HG}_{0,3}(\mathbf{r})$
$\mathbf{IG}^o_{3,3}(\mathbf{r},\varepsilon) = +\frac{c\cdot\mathbf{HG}_{2,1}(\mathbf{r}) - \sqrt{3}\cdot\mathbf{HG}_{0,3}(\mathbf{r})}{\sqrt{3+c^2}} = \frac{\sqrt{3}(c+1)\mathrm{Im}\mathbf{LG}_{0,3}(\mathbf{r}) - (c-3)\mathrm{Im}\mathbf{LG}_{1,1}(\mathbf{r})}{\sqrt{2}\cdot\sqrt{3+c^2}}$	$-\mathbf{HG}_{0,3}(\mathbf{r})$	$\sqrt{2}Im\mathbf{LG}_{0,3}(\mathbf{r})$	$+\mathbf{HG}_{2,1}(\mathbf{r})$
$\mathbf{IG}^o_{4,2}(\mathbf{r},\varepsilon) = -\frac{\mathbf{HG}_{3,1}(\mathbf{r}) + d\cdot\mathbf{HG}_{1,3}(\mathbf{r})}{\sqrt{1+d^2}} = \frac{(d-1)\mathrm{Im}\mathbf{LG}_{0,4}(\mathbf{r}) + (d+1)\mathrm{Im}\mathbf{LG}_{1,2}(\mathbf{r})}{\sqrt{1+d^2}}$	$-\mathbf{HG}_{3,1}(\mathbf{r})$	$\sqrt{2}Im\mathbf{LG}_{1,2}(\mathbf{r})$	$-\mathbf{HG}_{1,3}(\mathbf{r})$
$\mathbf{IG}^o_{4,4}(\mathbf{r},\varepsilon) = +\frac{d\cdot\mathbf{HG}_{3,1}(\mathbf{r}) - \mathbf{HG}_{1,3}(\mathbf{r})}{\sqrt{1+d^2}} = \frac{(d+1)\mathrm{Im}\mathbf{LG}_{0,4}(\mathbf{r}) + (d-1)\mathrm{Im}\mathbf{LG}_{1,2}(\mathbf{r})}{\sqrt{1+d^2}}$	$-\mathbf{HG}_{1,3}(\mathbf{r})$	$\sqrt{2}Im\mathbf{LG}_{0,4}(\mathbf{r})$	$+\mathbf{HG}_{3,1}(\mathbf{r})$

with

$$c_4 = \tfrac{4}{3}\sqrt{13+3\varepsilon^2}\cos\phi, \quad s_4 = \tfrac{4}{3}\sqrt{13+3\varepsilon^2}\sin\phi,$$
$$\phi = \tfrac{1}{3}\arccos\left(-\frac{35-18\varepsilon^2}{(13+3\varepsilon^2)^{3/2}}\right).$$

Asymptotic expansions of all three roots at small and large values of ε are written in Table 2.2

Table 2.2 Asymptotic expansions of the roots (2.18)

	$\varepsilon \approx 0$	$\varepsilon \to \infty$
λ_0^e	$-3\varepsilon^2 + \frac{33}{16}\varepsilon^4 - \dots$	$-4\varepsilon + 4 - \frac{6}{\varepsilon} + \frac{12}{\varepsilon^2} - \dots$
λ_1^e	$4 + \frac{8\varepsilon^2}{3} - \frac{56\varepsilon^4}{27} + \dots$	$12 - \frac{24}{\varepsilon^2} - \frac{24}{\varepsilon^4} + \dots$
λ_2^e	$16 + \frac{\varepsilon^2}{3} - \frac{5\varepsilon^4}{432} + \dots$	$4\varepsilon + 4 + \frac{6}{\varepsilon} + \frac{12}{\varepsilon^2} + \dots$

Table 2.3 Limiting cases of the normalized even IG modes at $p = 4$

	$\varepsilon \to -\infty$	$\varepsilon = 0$	$\varepsilon \to +\infty$
$\mathbf{IG}_{4,0}^e(\mathbf{r}, \varepsilon)$	$+\mathbf{HG}_{4,0}(\mathbf{r})$	$\mathbf{LG}_{2,0}(\mathbf{r})$	$+\mathbf{HG}_{0,4}(\mathbf{r})$
$\mathbf{IG}_{4,2}^e(\mathbf{r}, \varepsilon)$	$+\mathbf{HG}_{2,2}(\mathbf{r})$	$\sqrt{2}\mathrm{Re}\mathbf{LG}_{1,2}(\mathbf{r})$	$-\mathbf{HG}_{2,2}(\mathbf{r})$
$\mathbf{IG}_{4,4}^e(\mathbf{r}, \varepsilon)$	$+\mathbf{HG}_{0,4}(\mathbf{r})$	$\sqrt{2}\mathrm{Re}\mathbf{LG}_{0,4}(\mathbf{r})$	$+\mathbf{HG}_{4,0}(\mathbf{r})$

These formulae are very useful for investigating the limiting cases $\varepsilon \to 0$ and $\varepsilon \to \infty$.

For $p = 4$, the even, non-normalized IG modes are given by

$$
\begin{aligned}
\mathcal{I}_{4,2m}^e(\mathbf{r}, \varepsilon) &= 32\left\{2\mathrm{Re}\mathcal{L}_{0,4}(\mathbf{r}) - A_1\mathrm{Re}\mathcal{L}_{1,2}(\mathbf{r}) + A_0\mathcal{L}_{2,0}(\mathbf{r})\right\} \\
&= (A_0 + A_1 + 1)\mathcal{H}_{4,0}(\mathbf{r}) + (2A_0 - 6)\mathcal{H}_{2,2}(\mathbf{r}) \\
&\quad + (A_0 - A_1 + 1)\mathcal{H}_{0,4}(\mathbf{r}),
\end{aligned}
\tag{2.19}
$$

with $m = 0, 1, 2$ and

$$
\begin{aligned}
A_0 = A_0(m, \varepsilon) &= \frac{(\lambda_m^e - 4)(\lambda_m^e - 16)}{4\varepsilon^2} - 1, \\
A_1 = A_1(m, \varepsilon) &= \frac{\lambda_m^e - 16}{\varepsilon}.
\end{aligned}
\tag{2.20}
$$

We do not present the normalized version in order to avoid cumbersome fractions, and give the normalized IG modes only in the limiting cases (Table 2.3).

For numerical computation of the IG modes, when the ellipticity parameter is not very small and not very large, the Cardano formula along with expansions (2.19) should be employed, with adding the normalizing multiplier if necessary. If, however, the ellipticity parameter is close to the limiting values, then the asymptotic expressions of the characteristic roots are more convenient. For instance, if $m = 2$ and $\varepsilon \to 0$, we get $A_0 = -\frac{1}{144}\varepsilon^2 + O(\varepsilon^4)$, $A_1 = \frac{1}{3}\varepsilon + O(\varepsilon^3)$, and, thus, $\mathcal{I}_{4,4}^e(\mathbf{r}, \varepsilon) = 64\mathrm{Re}\mathcal{L}_{0,4}(\mathbf{r}) - \frac{32}{3}\varepsilon\mathrm{Re}\mathcal{L}_{1,2}(\mathbf{r}) + O(\varepsilon^2)$.

For many values of ε, the roots of the Eq. (2.17) can be easily found without using the Cardano formula, for example, for $\varepsilon = \left\{\frac{1}{2}\sqrt{5},\ 2,\ \frac{6}{5}\sqrt{5},\ \frac{3}{2}\sqrt{5},\ \sqrt{10},\ 2\sqrt{5}\right\}$. It is enough to choose some integer λ as a root and find which ε it corresponds to (relative to ε, Eq. (2.17) is quadratic). In particular, for $\varepsilon = 2$ we get $\lambda_0^e = 6 - 2\sqrt{33}$, $\lambda_1^e = 8$, $\lambda_2^e = 6 + 2\sqrt{33}$.

In addition, it can be shown that a simple interrelation exists between the IG modes constructed for the parameters ε and –ε:

$$\begin{aligned}\mathbf{IG}^{e}_{p,2m}(x, y, -\varepsilon) &= (-1)^{m}\mathbf{IG}^{e}_{p,2m}(y, x, \varepsilon),\\ \mathbf{IG}^{o}_{p,2m}(x, y, -\varepsilon) &= (-1)^{m+1}\mathbf{IG}^{o}_{p,2m}(y, x, \varepsilon),\\ \mathbf{IG}^{o}_{p,2m+1}(x, y, -\varepsilon) &= (-1)^{m}\mathbf{IG}^{e}_{p,2m+1}(y, x, \varepsilon).\end{aligned} \tag{2.21}$$

The first two formulae here are written for the even value of p, whereas the last one—for an odd p. This demonstrates once again that the case of negative ε is no less important in investigating the IG modes as the case of positive ε. In particular, Eq. (2.21) indicate that the IG modes with the negative parameter ε are also orthogonal to each other, similarly to the modes with the positive ε.

The proof of the relations (2.21) is based upon the following property of the characteristic polynomials:

$$\begin{aligned}\chi_p^{(e,o)}(\lambda, \varepsilon) &= \chi_p^{(e,o)}(\lambda, -\varepsilon), \quad \text{if } p \text{ is even},\\ \chi_p^{o}(\lambda, \varepsilon) &= \chi_p^{e}(\lambda, -\varepsilon), \quad \text{if } p \text{ is odd}.\end{aligned} \tag{2.22}$$

Now we consider the case $p = 5$. For the even IG modes, the characteristic equation is given by

$$\begin{aligned}\chi_5^{e}(\lambda, \varepsilon) = \lambda^3 &- (35 + 3\varepsilon)\lambda^2 + (259 + 102\varepsilon - 13\varepsilon^2)\lambda\\ &- 225 - 675\varepsilon + 205\varepsilon^2 + 15\varepsilon^3 = 0.\end{aligned} \tag{2.23}$$

We again apply the variant of the Cardano formula when the cubic equation has three real-valued roots, and these roots in the ascending order read as

$$\begin{aligned}&\lambda_0^{e} = \left(\tfrac{35}{3} + \varepsilon\right) - 2c_5, \quad \lambda_1^{e} = \left(\tfrac{35}{3} + \varepsilon\right) + c_5 - s_5\sqrt{3},\\ &\lambda_2^{e} = \left(\tfrac{35}{3} + \varepsilon\right) + c_5 + s_5\sqrt{3},\end{aligned} \tag{2.24}$$

where

$$\begin{aligned}&c_5 = \tfrac{4}{3}\sqrt{28 - 6\varepsilon + 3\varepsilon^2}\cos\phi, \quad s_5 = \tfrac{4}{3}\sqrt{28 - 6\varepsilon + 3\varepsilon^2}\sin\phi,\\ &\phi = \tfrac{1}{3}\arccos\left(-\frac{80 + 9\varepsilon - 18\varepsilon^2}{(28 - 6\varepsilon + 3\varepsilon^2)^{3/2}}\right)\end{aligned}$$

Asymptotic expansions of all three roots at small and large ε are given in Table 2.4. For $p = 5$, even IG modes (without normalization) are given by

$$\begin{aligned}\mathcal{I}^{e}_{5,2m+1}(\mathbf{r}, \varepsilon) &= 64\sqrt{2}\left\{2\mathrm{Re}\mathcal{L}_{0,5}(\mathbf{r}) - A_1\mathrm{Re}\mathcal{L}_{1,3}(\mathbf{r}) + A_0\mathrm{Re}\mathcal{L}_{2,1}(\mathbf{r})\right\}\\ &= (A_0 + A_1 + 1)\mathcal{H}_{5,0}(\mathbf{r}) + 2(A_0 - A_1 - 5)\mathcal{H}_{3,2}(\mathbf{r})\\ &\quad + (A_0 - 3A_1 + 5)\mathcal{H}_{1,4}(\mathbf{r}),\end{aligned} \tag{2.25}$$

Table 2.4 Asymptotic expansions of the roots (2.24)

	$\varepsilon \approx 0$	$\varepsilon \to \infty$
λ_0^e	$1 + 3\varepsilon - \varepsilon^2 + \ldots$	$-3\varepsilon + 13 - \frac{18}{\varepsilon} + \frac{18}{\varepsilon^2} + \ldots$
λ_1^e	$9 + \frac{11\varepsilon^2}{16} + \frac{3\varepsilon^3}{8} + \ldots$	$\varepsilon + 17 + \frac{8}{\varepsilon} - \frac{48}{\varepsilon^2} + \ldots$
λ_2^e	$25 + \frac{5\varepsilon^2}{16} + \ldots$	$5\varepsilon + 5 + \frac{10}{\varepsilon} + \frac{30}{\varepsilon^2} + \ldots$

with $m = 0, 1, 2$ and

$$A_0 = A_0(m, \varepsilon) = \frac{(\lambda_m^e - 9)(\lambda_m^e - 25)}{2\varepsilon^2} - \frac{5}{2},$$
$$A_1 = A_1(m, \varepsilon) = \frac{\lambda_m^e - 25}{\varepsilon}.$$

In the limiting cases, the normalized IG modes are shown in Table 2.5.

Equation (2.22) leads to relations between the roots of the characteristic polynomials: $\lambda_m^o(\varepsilon) = \lambda_m^e(-\varepsilon)$. Therefore, due to Eq. (2.21), odd IG modes are obtained from the even IG modes by swapping the indices of the HG modes, replacing ε by $-\varepsilon$, and adding a multiplier $(-1)^m$:

$$\begin{aligned}\mathcal{I}_{5,2m+1}^o(\mathbf{r}, \varepsilon) &= 64\sqrt{2}\left\{2\mathrm{Im}\mathcal{L}_{0,5}(\mathbf{r}) - B_1 Im\mathcal{L}_{1,3}(\mathbf{r}) + B_0\mathrm{Im}\mathcal{L}_{2,1}(\mathbf{r})\right\} \\ &= (B_0 + 3B_1 + 5)\mathcal{H}_{4,1}(\mathbf{r}) + 2(B_0 + B_1 - 5)\mathcal{H}_{2,3}(\mathbf{r}) \\ &\quad + (B_0 - B_1 + 1)\mathcal{H}_{0,5}(\mathbf{r}),\end{aligned} \tag{2.26}$$

with $m = 0, 1, 2$ and

$$B_0 = B_0(m, \varepsilon) = \frac{(\lambda_m^o - 9)(\lambda_m^o - 25)}{2\varepsilon^2} - \frac{5}{2},$$
$$B_1 = B_1(m, \varepsilon) = \frac{\lambda_m^o - 25}{\varepsilon}.$$

Similarly to the above considered case $p = 4$, for many values of ε, the roots of the characteristic polynomials can be found without using the Cardano formula. For instance, for the even IG modes, the following values can be chosen: $\varepsilon = \left\{1, \frac{3}{2}, 3, 5, \frac{21}{4}, 8\right\}$, whereas for the odd IG modes—the values $\varepsilon = \left\{\frac{3}{4}, \frac{7}{5}, \frac{7}{3}, 3, \frac{88}{15}, \frac{45}{4}\right\}$. In particular, for $\varepsilon = 3$, we get the roots $\lambda_m^e = \{0, 16, 28\}$

Table 2.5 Limiting cases of normalized even IG modes at $p = 5$

	$\varepsilon \to -\infty$	$\varepsilon = 0$	$\varepsilon \to +\infty$
$\mathbf{IG}_{5,1}^e(\mathbf{r}, \varepsilon)$	$+\mathbf{HG}_{5,0}(\mathbf{r})$	$\sqrt{2}\mathrm{Re}\mathbf{LG}_{2,1}(\mathbf{r})$	$+\mathbf{HG}_{1,4}(\mathbf{r})$
$\mathbf{IG}_{5,3}^e(\mathbf{r}, \varepsilon)$	$+\mathbf{HG}_{3,2}(\mathbf{r})$	$\sqrt{2}\mathrm{Re}\mathbf{LG}_{1,3}(\mathbf{r})$	$-\mathbf{HG}_{3,2}(\mathbf{r})$
$\mathbf{IG}_{5,5}^e(\mathbf{r}, \varepsilon)$	$+\mathbf{HG}_{1,4}(\mathbf{r})$	$\sqrt{2}\mathrm{Re}\mathbf{LG}_{0,5}(\mathbf{r})$	$+\mathbf{HG}_{5,0}(\mathbf{r})$

and $\lambda_m^o = \left\{8 - 2\sqrt{97},\ 10,\ 8 + 2\sqrt{97}\right\}$. Therefore, for example,

$$\begin{aligned} \mathcal{I}_{5,1}^e(\mathbf{r}, 3) &= 48\{\mathcal{H}_{5,0}(\mathbf{r}) + 10\mathcal{H}_{3,2}(\mathbf{r}) + 15\mathcal{H}_{1,4}(\mathbf{r})\}, \\ \mathcal{I}_{5,3}^o(\mathbf{r}, 3) &= 48\{-\mathcal{H}_{0,5}(\mathbf{r}) + 10\mathcal{H}_{2,3}(\mathbf{r}) + 5\mathcal{H}_{4,1}(\mathbf{r})\}. \end{aligned} \tag{2.27}$$

There is one more case left with the cubic characteristic equation. This is the case $p = 6$ for the odd IG modes:

$$\chi_6^o(\lambda, \varepsilon) = \lambda^3 - 56\lambda^2 + 16(49 - \varepsilon^2)\lambda - 384(6 - \varepsilon^2) = 0. \tag{2.28}$$

Then

$$\begin{aligned} \lambda_0^e &= \tfrac{56}{3} - 2c_6, \quad \lambda_1^e = \tfrac{56}{3} + c_6 - \sqrt{3}s_6, \\ \lambda_2^e &= \tfrac{56}{3} + c_6 + \sqrt{3}s_6, \end{aligned} \tag{2.29}$$

with

$$\begin{aligned} c_6 &= \tfrac{4}{3}\sqrt{49 + 3\varepsilon^2}\cos\phi, \quad s_6 = \tfrac{4}{3}\sqrt{49 + 3\varepsilon^2}\sin\phi, \\ \phi &= \tfrac{1}{3}\arccos\left(-\frac{143 - 18\varepsilon^2}{(49 + 3\varepsilon^2)^{3/2}}\right). \end{aligned}$$

Correspondingly, the odd, non-normalized IG modes are given by

$$\begin{aligned} \mathcal{I}_{6,2m+2}^o(\mathbf{r}, \varepsilon) &= 128\{4\mathrm{Im}\mathcal{L}_{0,6}(\mathbf{r}) - 2B_2\mathrm{Im}\mathcal{L}_{1,4}(\mathbf{r}) + 2B_1\mathrm{Im}\mathcal{L}_{2,2}(\mathbf{r})\} \\ &= 2(B_1 + 2B_2 + 3)\mathcal{H}_{5,1}(\mathbf{r}) + 4(B_1 - 5)\mathcal{H}_{3,3}(\mathbf{r}) \\ &\quad + 2(B_1 - 2B_2 + 3)\mathcal{H}_{1,5}(\mathbf{r}), \end{aligned} \tag{2.30}$$

with $m = 0, 1, 2$ and

$$B_1 = \frac{(\lambda_m^e - 16)(\lambda_m^e - 36)}{2\varepsilon^2} - 3, \quad B_2 = \frac{\lambda_m^e - 36}{\varepsilon}.$$

Cubic Eq. (2.28) can be solved without using the Cardano formula, for instance, for $\varepsilon = \left\{\tfrac{2}{3}\sqrt{10},\ \sqrt{6},\ \tfrac{3}{2}\sqrt{6}, 4, \tfrac{3}{2}\sqrt{11}, 2\sqrt{7}, 2\sqrt{11}\right\}$. In particular, for $\varepsilon = \sqrt{6}$, we get the roots $\lambda_m^o = \left\{0,\ 28 - 4\sqrt{6},\ 28 + 4\sqrt{6}\right\}$. Therefore,

$$\begin{aligned} \mathcal{I}_{6,2}^o(\mathbf{r}, \sqrt{6}) &= 24\left\{(4 - \sqrt{6})\mathcal{H}_{5,1}(\mathbf{r}) + \tfrac{20}{3}\mathcal{H}_{3,3}(\mathbf{r}) + (4 + \sqrt{6})\mathcal{H}_{1,5}(\mathbf{r})\right\}, \\ \mathcal{I}_{6,4}^o(\mathbf{r}, \sqrt{6}) &= \tfrac{8}{3}(6 + \sqrt{6})\left\{-\tfrac{1}{5}(3 + 2\sqrt{6})\mathcal{H}_{5,1}(\mathbf{r}) - 2\mathcal{H}_{3,3}(\mathbf{r}) + \mathcal{H}_{1,5}(\mathbf{r})\right\}, \\ \mathcal{I}_{6,6}^o(\mathbf{r}, \sqrt{6}) &= \tfrac{8}{3}(6 - \sqrt{6})\left\{\mathcal{H}_{5,1}(\mathbf{r}) - 2\mathcal{H}_{3,3}(\mathbf{r}) - \tfrac{1}{5}(3 - 2\sqrt{6})\mathcal{H}_{1,5}(\mathbf{r})\right\}. \end{aligned} \tag{2.31}$$

Even IG Modes at P = 6

Now we consider the case $p = 6$ for the even IG modes. The characteristic polynomial is quartic:

$$\chi_6^e(\lambda, \varepsilon) = \lambda^4 - 56\lambda^3 + 8(98 - 5\varepsilon^2)\lambda^2 - 96(24 - 17\varepsilon^2)\lambda - 144\varepsilon^2(96 - \varepsilon^2) = 0. \tag{2.32}$$

Its roots $\lambda_m^e(\varepsilon)$ can be obtained by the Ferrari formula. We use the variant of this formula presented in [19]. Let us introduce an auxiliary variable $h = \frac{4}{3}(49 + 5\varepsilon^2) + C_6 + S_6\sqrt{3}$ with

$$C_6 = \tfrac{1}{3}\sqrt{889 + 178\varepsilon^2 + 13\varepsilon^4}\cos\phi,$$
$$S_6 = \tfrac{1}{3}\sqrt{889 + 178\varepsilon^2 + 13\varepsilon^4}\sin\phi,$$
$$\phi = \tfrac{1}{3}\arccos\left(-\frac{24013 - 6033\varepsilon^2 + 1317\varepsilon^4 + 35\varepsilon^6}{(889 + 178\varepsilon^2 + 13\varepsilon^4)^{3/2}}\right).$$

Then four roots of Eq. (2.32) in the ascending order are expressed via h as follows:

$$\lambda_0^e = 14 - \sqrt{h} - \sqrt{4(49 + 5\varepsilon^2) - h - \frac{64(9 - 2\varepsilon^2)}{\sqrt{h}}},$$
$$\lambda_1^e = 14 - \sqrt{h} + \sqrt{4(49 + 5\varepsilon^2) - h - \frac{64(9 - 2\varepsilon^2)}{\sqrt{h}}},$$
$$\lambda_2^e = 14 + \sqrt{h} - \sqrt{4(49 + 5\varepsilon^2) - h + \frac{64(9 - 2\varepsilon^2)}{\sqrt{h}}},$$
$$\lambda_3^e = 14 + \sqrt{h} + \sqrt{4(49 + 5\varepsilon^2) - h + \frac{64(9 - 2\varepsilon^2)}{\sqrt{h}}}. \tag{2.33}$$

Asymptotic expansions of all roots at small and large ε are written in Table 2.6.

Then the non-normalized IG modes are expressed via the LG and HG modes in the following way:

Table 2.6 Asymptotic expansions of the roots (2.33)

	$\varepsilon \approx 0$	$\varepsilon \to \infty$
λ_0^e	$-6\varepsilon^2 + \frac{129\varepsilon^4}{16} - \ldots$	$-6\varepsilon + 6 - \frac{15}{\varepsilon} + \frac{60}{\varepsilon^2} - \ldots$
λ_1^e	$4 + \frac{31\varepsilon^2}{6} - \frac{28069\varepsilon^4}{3456} + \ldots$	$-2\varepsilon + 22 - \frac{21}{\varepsilon} - \frac{60}{\varepsilon^2} + \ldots$
λ_2^e	$16 + \frac{8\varepsilon^2}{15} + \frac{3193\varepsilon^4}{54000} - \ldots$	$2\varepsilon + 22 + \frac{21}{\varepsilon} - \frac{60}{\varepsilon^2} - \ldots$
λ_3^e	$36 + \frac{3\varepsilon^2}{10} + \frac{3\varepsilon^4}{16000} + \ldots$	$6\varepsilon + 6 + \frac{15}{\varepsilon} + \frac{60}{\varepsilon^2} + \ldots$

Table 2.7 Limiting cases of normalized even IG modes at $p = 6$

	$\varepsilon \to -\infty$	$\varepsilon = 0$	$\varepsilon \to +\infty$
$\mathbf{IG}^e_{6,0}(\mathbf{r}, \varepsilon)$	$-\mathbf{HG}_{6,0}(\mathbf{r})$	$\mathbf{LG}_{3,0}(\mathbf{r})$	$-\mathbf{HG}_{0,6}(\mathbf{r})$
$\mathbf{IG}^e_{6,2}(\mathbf{r}, \varepsilon)$	$-\mathbf{HG}_{4,2}(\mathbf{r})$	$\sqrt{2}\mathrm{Re}\mathbf{LG}_{2,2}(\mathbf{r})$	$+\mathbf{HG}_{2,4}(\mathbf{r})$
$\mathbf{IG}^e_{6,4}(\mathbf{r}, \varepsilon)$	$-\mathbf{HG}_{2,4}(\mathbf{r})$	$\sqrt{2}\mathrm{Re}\mathbf{LG}_{1,4}(\mathbf{r})$	$-\mathbf{HG}_{4,2}(\mathbf{r})$
$\mathbf{IG}^e_{6,6}(\mathbf{r}, \varepsilon)$	$-\mathbf{HG}_{0,6}(\mathbf{r})$	$\sqrt{2}\mathrm{Re}\mathbf{LG}_{0,6}(\mathbf{r})$	$+\mathbf{HG}_{6,0}(\mathbf{r})$

$$
\begin{aligned}
\mathcal{I}^e_{6,2m}(\mathbf{r}, \varepsilon) &= 128\big\{4\mathrm{Re}\mathcal{L}_{0,6}(\mathbf{r}) - 2A_2\mathrm{Re}\mathcal{L}_{1,4}(\mathbf{r}) \\
&\quad +2A_1\mathrm{Re}\mathcal{L}_{2,2}(\mathbf{r}) - 3A_0\mathcal{L}_{3,0}(\mathbf{r})\big\} \\
&= (A_0 + A_1 + A_2 + 1)\mathcal{H}_{6,0}(\mathbf{r}) \\
&\quad + (3A_0 + A_1 - 5A_2 - 15)\mathcal{H}_{4,2}(\mathbf{r}) \\
&\quad + (3A_0 - A_1 - 5A_2 + 15)\mathcal{H}_{2,4}(\mathbf{r}) \\
&\quad + (A_0 - A_1 + A_2 - 1)\mathcal{H}_{0,6}(\mathbf{r}),
\end{aligned}
\tag{2.34}
$$

with $m = 0, 1, 2, 3$ and

$$
\begin{aligned}
A_0 &= \frac{(\lambda^e_m - 4)(\lambda^e_m - 16)(\lambda^e_m - 36)}{12\varepsilon^3} - \frac{4(\lambda^e_m - 24)}{3\varepsilon}, \\
A_1 &= \frac{(\lambda^e_m - 16)(\lambda^e_m - 36)}{2\varepsilon^2} - 3, \quad A_2 = \frac{\lambda^e_m - 36}{\varepsilon}.
\end{aligned}
$$

In the limiting cases, the normalized IG modes are shown in Table 2.7.

It is obvious that when the index p and, respectively, degree of the characteristic polynomial, increase, finding all its roots in a simple form and without numerical methods becomes less and less possible. Nevertheless, it is easy to check that for the value $\varepsilon = \sqrt{19}$ one gets the roots $\lambda^e_m = \left\{-22, 18 - 4\sqrt{6}, 18 + 4\sqrt{6}, 42\right\}$, whereas the value $\varepsilon = 3\sqrt{3}$ leads to the roots $\lambda^e_m = \left\{2 - 8\sqrt{13}, 26 - 4\sqrt{22}, 2 + 8\sqrt{13}, 26 + 4\sqrt{22}\right\}$. Thus, for example,

$$
\begin{aligned}
\mathcal{I}^e_{6,0}(\mathbf{r}, \sqrt{19}) &= 8\left\{\left(7 - \frac{31}{\sqrt{19}}\right)\mathcal{H}_{6,0}(\mathbf{r}) + \left(5 - \frac{35}{\sqrt{19}}\right)\mathcal{H}_{4,2}(\mathbf{r})\right. \\
&\quad \left. - \left(5 + \frac{35}{\sqrt{19}}\right)\mathcal{H}_{2,4}(\mathbf{r}) - \left(7 + \frac{31}{\sqrt{19}}\right)\mathcal{H}_{0,6}(\mathbf{r})\right\}.
\end{aligned}
\tag{2.35}
$$

2.1.3 Numerical Methods and IG Modes

It is known that for all characteristic polynomials $\chi_p^{(e,o)}(\lambda, \varepsilon)$, regardless the index p and the parity flag (e, o), the coefficient at any term of the form $\lambda^\nu \varepsilon^\mu$ is an integer number. Therefore, the series expansions of the roots $\lambda_m^{(e,o)}(\varepsilon)$ by powers of ε (both at $\varepsilon \to 0$ and at $\varepsilon \to \pm\infty$) have the expansion coefficients which are rational numbers (see the above formulae as examples of such expansions). Consequently, asymptotic expansions of the IG modes by the LG and HG modes have the coefficients that can be represented as series of the powers of ε (if $\varepsilon \to 0$, then by powers ε, ε^2, …; if $\varepsilon \to \pm\infty$, then by the negative powers), and the coefficients in these series are also rational ones.

For instance, let us consider the mode $\mathbf{IG}_{3,1}^e(\mathbf{r}, \varepsilon)$, writing it in the following form:

$$\mathbf{IG}_{3,1}^e(\mathbf{r}, \varepsilon) = \frac{1}{2\sqrt{2\pi}} \cdot \frac{(A_0 + 1)\mathcal{H}_{3,0}(\mathbf{r}) + (A_0 - 3)\mathcal{H}_{1,2}(\mathbf{r})}{\sqrt{(A_0 + 1)^2 3 + (A_0 - 3)^2}}. \tag{2.36}$$

Here $2\sqrt{2\pi}$ [illegible] $\|\mathcal{H}_{1,2}(\mathbf{r})\| =$ [illegible] $\|\mathcal{L}_{1,1}(\mathbf{r})\|$ and

$$A_0 = \frac{\lambda_0^e - 9}{\varepsilon} = \frac{-4 + \varepsilon - 2\sqrt{4 - 2\varepsilon + \varepsilon^2}}{\varepsilon}.$$

If in the expansion coefficients at $\varepsilon \to 0$ and $\varepsilon \to +\infty$ we retain only terms up to ε^3 and ε^{-4} respectively, then we get the following asymptotics.

If $\varepsilon \to 0$ then

$$\begin{aligned}\mathbf{IG}_{3,1}^e(\mathbf{r}, \varepsilon) = \frac{1}{2\sqrt{2\pi}} \cdot \Bigg\{&\left(-\frac{1}{2} + \frac{\varepsilon}{16} + \frac{7\varepsilon^2}{256} + \frac{11\varepsilon^3}{2048} - \ldots\right)\mathcal{H}_{3,0}(\mathbf{r}) \\ &+ \left(-\frac{1}{2} - \frac{3\varepsilon}{16} - \frac{9\varepsilon^2}{256} + \frac{15\varepsilon^3}{2048} + \ldots\right)\mathcal{H}_{1,2}(\mathbf{r})\Bigg\}.\end{aligned} \tag{2.37}$$

In particular, since $-\frac{1}{2}\{\mathcal{H}_{3,0}(\mathbf{r}) + \mathcal{H}_{1,2}(\mathbf{r})\} = 4\sqrt{2}\mathrm{Re}\mathcal{L}_{1,1}(\mathbf{r})$, then, as a limiting case, we get an identity that is already well known: $\mathbf{IG}_{3,1}^e(\mathbf{r}, 0) = \sqrt{2}\mathrm{Re}\mathbf{LG}_{1,1}(\mathbf{r})$.

If $\varepsilon \to +\infty$ then

$$\begin{aligned}\mathbf{IG}_{3,1}^e(\mathbf{r}, \varepsilon) = \frac{1}{2\sqrt{2\pi}}\Bigg\{&\left(-\frac{1}{2\varepsilon} - \frac{1}{2\varepsilon^2} + \frac{1}{16\varepsilon^3} + \frac{19}{16\varepsilon^4} + \ldots\right)\mathcal{H}_{3,0}(\mathbf{r}) \\ &+ \left(-1 + \frac{3}{8\varepsilon^2} + \frac{3}{4\varepsilon^3} + \frac{45}{128\varepsilon^4} + \ldots\right)\mathcal{H}_{1,2}(\mathbf{r})\Bigg\}.\end{aligned} \tag{2.38}$$

Here, in the limiting case, we get a HG mode: $\mathbf{IG}_{3,1}^e(\mathbf{r}, \infty) = -\mathbf{HG}_{1,2}(\mathbf{r})$.

2.1.4 Applying the Padé Approximants for Approximate Computation of IG Modes

In this Section, we demonstrate on a concrete example how the Padé approximants can be employed for computing the IG modes. In the limiting case, IG mode reduces to a HG mode: $\mathbf{IG}^e_{3,1}(\mathbf{r},\infty) = -\mathbf{HG}_{1,2}(\mathbf{r})$ (third row in Table 2.1). To obtain an approximate expansion of the mode $\mathbf{IG}^e_{3,1}(\mathbf{r},\varepsilon)$ by the HG modes, which is suitable for the whole range $\varepsilon \in [0,\infty)$, we use the Padé approximants [20]. The Padé approximant of some function $f(z)$ is given by the expression

$$f(z|[L/M]) = \frac{a_0 + a_1 z + \cdots + a_L z^L}{1 + b_1 z + \cdots + b_M z^M}, \tag{2.39}$$

where the coefficients a_k and b_k are chosen so that at small values z, it has several first terms of the Taylor expansion by the powers of z, exactly the same as those of the function $f(z)$, whereas at large z, several first terms of the asymptotic expansion by the powers $(1/z)$ are the same as those of the function $f(z)$.

Combining both formulae (2.37) and (2.38) into the one, we get

$$\mathbf{IG}^e_{3,1}(\mathbf{r},\varepsilon) = \frac{1}{2\sqrt{2\pi}} \cdot \left\{C_{3,0}(\varepsilon)\mathcal{H}_{3,0}(\mathbf{r}) + C_{1,2}(\varepsilon)\mathcal{H}_{1,2}(\mathbf{r})\right\}, \tag{2.40}$$

where exact representations of the functions of ε are given by Eq. (2.36), whereas their approximations in the form of Eq. (2.39) depend on the choice of the parameters L and M. Since

$$C_{3,0}(0) = C_{1,2}(0) = -\frac{1}{2}, \quad C_{3,0}(\varepsilon) \underset{\varepsilon\to\infty}{\approx} -\frac{1}{2\varepsilon}, \quad C_{1,2}(\infty) = -1,$$

then the Padé approximants of the functions $C_{3,0}(\varepsilon)$ and $C_{1,2}(\varepsilon)$ should have the following form:

$$\begin{aligned} C_{3,0}(\varepsilon|[L/(L+1)]) &= \frac{-\frac{1}{2} + a_1\varepsilon + \cdots + a_L\varepsilon^L}{1 + b_1\varepsilon + \cdots + b_L\varepsilon^L - 2a_L\varepsilon^{L+1}}, \\ C_{1,2}(\varepsilon|[L/L]) &= \frac{-\frac{1}{2} + \tilde{a}_1\varepsilon + \cdots + \tilde{a}_L\varepsilon^L}{1 + \tilde{b}_1\varepsilon + \cdots + \tilde{b}_{L-1}\varepsilon^{L-1} - \tilde{a}_L\varepsilon^L}. \end{aligned} \tag{2.41}$$

For each fixed value of the parameter L, the coefficients of the rational functions in Eq. (2.41) can be found by solving systems of equations. These equations are obtained if we expand functions (2.41) into series in powers of ε at $\varepsilon \to 0$ and in powers of $1/\varepsilon$ at $\varepsilon \to +\infty$, and then make equal these expansions to those already known from formulae (2.37) and (2.38). This approach is easy to implement in modern computer algebra systems. We here give only the simplest results when the obtained fractions are not very cumbersome yet:

$$C_{3,0}(\varepsilon|[2/3]) = \frac{-\frac{1}{2} - \frac{1711}{6176}\varepsilon - \frac{20}{193}\varepsilon^2}{1 + \frac{2097}{3088}\varepsilon + \frac{1071}{3088}\varepsilon^2 + \frac{40}{193}\varepsilon^3},$$
$$C_{1,2}(\varepsilon|[2/2]) = \frac{-\frac{1}{2} - \frac{3}{8}\varepsilon - \frac{27}{128}\varepsilon^2}{1 + \frac{3}{8}\varepsilon + \frac{27}{128}\varepsilon^2},$$
$$C_{1,2}(\varepsilon|[3/3]) = \frac{-\frac{1}{2} - \frac{2865}{6896}\varepsilon - \frac{13311}{55168}\varepsilon^2 - \frac{93}{862}\varepsilon^3}{1 + \frac{393}{862}\varepsilon + \frac{13311}{55168}\varepsilon^2 + \frac{93}{862}\varepsilon^3}. \tag{2.42}$$

Differences between the exact functions $C_{3,0}(\varepsilon)$ and $C_{1,2}(\varepsilon)$ and their Padé approximants are depicted in Fig. 2.1. In particular, the maximal difference $|C_{3,0}(\varepsilon) - C_{3,0}(\varepsilon|[L/(L+1)])|$ is 0.0187 for $L = 2$ and 0.0063 for $L = 3$, while the maximal difference $|C_{1,2}(\varepsilon) - C_{1,2}(\varepsilon|[L/L])|$ is 0.0631 for $L = 2$ and 0.0224 for $L = 3$. Here, the expansions $C_{3,0}(\varepsilon|[L/(L+1)])$ coincide with the expansions $C_{3,0}(\varepsilon)$ up to the terms ε^L and $1/\varepsilon^{L+1}$, inclusively, whereas for $C_{1,2}(\varepsilon|[L/L])$ and for $C_{1,2}(\varepsilon)$ the coincidence is up to the terms ε^L and $1/\varepsilon^{L-1}$. Increasing the parameter L makes it possible to obtain more and more accurate approximations of the IG mode, although it is accompanied by more cumbersome fractional rational expressions for the coefficients.

It is worth noting that using the Padé approximants does not require knowing the exact expressions for the functions $C_{3,0}(\varepsilon)$ and $C_{1,2}(\varepsilon)$. Instead, it quite suffices to know the first few terms of the series expansions of these functions (in powers of ε at $\varepsilon \to 0$ and in powers of $1/\varepsilon$ at $\varepsilon \to +\infty$). These expansions are obtained from the expansions of the roots of the characteristic equation, and for expanding the roots by the powers of ε or $1/\varepsilon$, well known conventional methods are used.

In this Section, the use of the Padé approximants is demonstrated for the mode $\mathbf{IG}^e_{3,1}(\mathbf{r}, \varepsilon)$, expanded into the series of HG modes. It is easy to see that this approach is suitable for any other IG modes. Its value for numerical computations of the modes $\mathbf{IG}^{(e,o)}_{p,q}(\mathbf{r}, \varepsilon)$ becomes obvious for large values of the index p because in this case roots of the characteristic equation cannot be expressed in radicals explicitly.

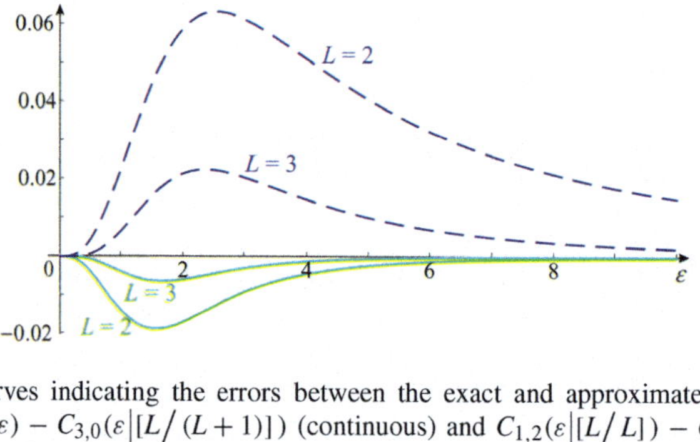

Fig. 2.1 The curves indicating the errors between the exact and approximate functions from Eq. (2.40): $C_{3,0}(\varepsilon) - C_{3,0}(\varepsilon|[L/(L+1)])$ (continuous) and $C_{1,2}(\varepsilon|[L/L]) - C_{1,2}(\varepsilon)$ (dashed) for $L = 2$ and $L = 3$. We intentionally took different signs for both expressions so that the curves locate on opposite sides of the x-axis

2.1.5 Numerical Simulation

In this Section, we obtain intensity distributions of the IG beams of different orders and with different values of the parameter ε, computed by solving the characteristic equation and by expansion into the LG modes. For computations, expressions (2.19), (2.25), (2.26), (2.30) and (2.34) are used, as well as Table 2.1.

Computation of the Ince-Gaussian Beams at P = 3

Figure 2.2 illustrates intensity distributions of the IG modes at $p = 3$ for several values of the parameter ε.

Figure 2.2 confirms that if the parameter ε is small ($\varepsilon \approx 0$), the mode $\mathbf{IG}^{e}_{3,3}$ reduces to $\mathrm{Re}\mathbf{LG}_{0,3}$ with the amplitude being proportional to cos 3φ, i.e. the intensity contains six petals, residing on a single light ring. In addition, Fig. 2.2 confirms that the intensity of the odd mode $\mathbf{IG}^{o}_{3,3}$ also contains six petals, but rotated by the angle of π/6.

As is also seen in Fig. 2.2, the intensity pattern of the even IG beam $\mathbf{IG}^{e}_{3,1}$ with a small parameter ε has two petals and a ring split into two arcs elongated along the vertical axis. Correspondingly, the intensity of the odd beam $\mathbf{IG}^{o}_{3,1}$ with a small parameter ($\varepsilon \approx 0$) has also two petals that divide two light rings, but directed along the horizontal axis.

At large values of the parameter ε ($\varepsilon \gg 1$), IG modes are defined by the HG modes and have Cartesian symmetry. It is seen that at $\varepsilon \gg 1$, the even mode $\mathbf{IG}^{o}_{3,1}$

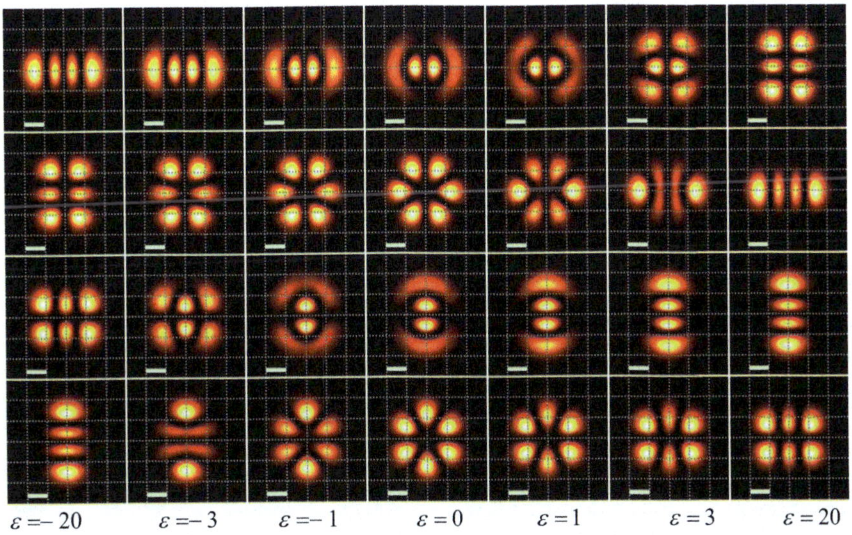

Fig. 2.2 Intensity distributions of the IG modes $\mathbf{IG}^{e}_{3,1}(\mathbf{r}, \varepsilon)$, $\mathbf{IG}^{e}_{3,3}(\mathbf{r}, \varepsilon)$, $\mathbf{IG}^{o}_{3,1}(\mathbf{r}, \varepsilon)$, and $\mathbf{IG}^{o}_{3,3}(\mathbf{r}, \varepsilon)$ (from top to bottom) for some values of the parameter ε. Other parameters are the following: wavelength $\lambda = 532$ nm, Gaussian beam wais radius $w_0 = 1$ mm

coincides by the intensity with the HG mode $\mathbf{HG}_{1,2}$ and has 6 local maxima, whereas the mode $\mathbf{IG}^e_{3,3}$ coincides with the HG mode $\mathbf{HG}_{3,0}$ and has 4 petals, residing on the horizontal axis. On the contrary, for the odd modes at $\varepsilon \gg 1$, the mode $\mathbf{IG}^o_{3,1}$ has 4 petals on the vertical axis, while the mode $\mathbf{IG}^o_{3,3}$ has 6 petals separated by two vertical and one horizontal zero-intensity lines.

Computation of the IG Beams at P = 4

Shown in Fig. 2.3 are intensity distributions of the IG beams with $p = 4$ for several values of the parameter ε.

As seen in Fig. 2.3, indeed, when the parameter ε is small, IG modes coincide with corresponding LG modes: the beam $\mathbf{IG}^e_{4,0}$ has the intensity with one central light spot and two surrounding light rings, the beam $\mathbf{IG}^e_{4,2}$ has the intensity with four petals on each of these two rings. The intensity of the beam $\mathbf{IG}^e_{4,4}$ has 8 petals residing on a ring, whereas the intensity of the beam $\mathbf{IG}^o_{4,2}$ also contains 8 petals, but they reside by 4 petals on each of the two rings. The beam $\mathbf{IG}^o_{4,4}$ has the intensity with 8 petals residing on a single ring.

Besides, Fig. 2.3 confirms that for the large values of the parameter ε, IG beams coincide with the HG beams. Therefore, intensity patterns of the IG beams at $\varepsilon \gg 1$ coincide with respective HG beams. The intensity of the beam $\mathbf{IG}^e_{4,0}$ has 5 petals,

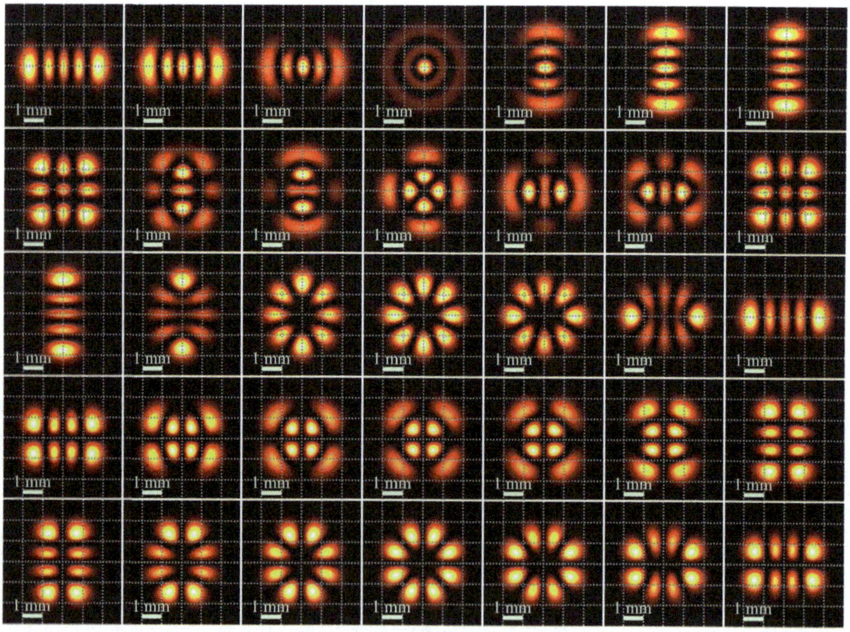

$\varepsilon = -20$ $\varepsilon = -3$ $\varepsilon = -1$ $\varepsilon = 0$ $\varepsilon = 1$ $\varepsilon = 3$ $\varepsilon = 20$

Fig. 2.3 Intensity distributions of the IG modes $\mathbf{IG}^e_{4,0}(\mathbf{r}, \varepsilon)$, $\mathbf{IG}^e_{4,2}(\mathbf{r}, \varepsilon)$, $\mathbf{IG}^e_{4,4}(\mathbf{r}, \varepsilon)$, $\mathbf{IG}^o_{4,2}(\mathbf{r}, \varepsilon)$, and $\mathbf{IG}^o_{4,4}(\mathbf{r}, \varepsilon)$ (from top to bottom) for some values of the parameter ε. Other parameters are the following: wavelength $\lambda = 532$ nm, Gaussian beam wais radius $w_0 = 1$ mm

residing on the vertical axis and separated by four horizontal zero-intensity lines; the intensity of the beam $\mathbf{IG}^{e}_{4,2}$ has 9 petals, separated by two horizontal and two vertical zero-intensity lines; the intensity of the beam $\mathbf{IG}^{e}_{4,4}$ has 5 petals, residing on the horizontal axis and separated by four vertical zero-intensity lines; the beam $\mathbf{IG}^{o}_{4,2}$ has the intensity with 8 petals, separated by one vertical and three horizontal zero-intensity lines, and the beam $\mathbf{IG}^{o}_{4,4}$ has the same intensity as that of the beam $\mathbf{IG}^{o}_{4,2}$, but rotated by $\pi/2$.

A comparison of Fig. 2.2 for $p = 3$ and Fig. 2.3 for $p = 4$ reveals that if p is even then changing the sign of the parameter ε leads to rotation of the intensity pattern (Fig. 2.3), whereas if p is odd then changing the sign of ε leads not only to rotation of the intensity pattern, but it is also changed (Fig. 2.2). This follows from Eq. (2.21). This equation indicates that for even value p, the IG functions are not changed themselves. Instead, only their arguments are swapped: $\mathbf{IG}^{e}_{p,2m}(x, y, -\varepsilon) = (-1)^m \mathbf{IG}^{e}_{p,2m}(y, x, \varepsilon)$ or $\mathbf{IG}^{o}_{p,2m}(x, y, -\varepsilon) = (-1)^{m+1} \mathbf{IG}^{o}_{p,2m}(y, x, \varepsilon)$. Due to this arguments interchange, the intensity rotates by $\pi/2$ (Fig. 2.3). On the contrary, for odd values p, when ε changes sign, not only the arguments are swapped, but the functions are changed (an even to an odd, or vice versa): $\mathbf{IG}^{o}_{p,2m+1}(x, y, -\varepsilon) = (-1)^m \mathbf{IG}^{e}_{p,2m+1}(y, x, \varepsilon)$.

Computation of the IG Beams at P = 5

Figure 2.4 depicts the intensity distributions of the IG beams with $p = 5$ for several values of the parameter ε.

Similarly to the smaller values p, Fig. 2.4 demonstrates that if the parameter ε is small then the IG modes coincide with the real and imaginary parts of the LG modes, and, on the contrary, if the parameter ε is large by modulus, the IG modes reduce to one of the HG modes.

Computation of the IG Beams at P = 6

Figure 2.5 illustrates the intensity distributions of the IG beams with $p = 6$ for several values of the parameter ε.

Figure 2.5 confirms that if the ellipticity parameter ε is small then the mode $\mathbf{IG}^{e}_{6,0}$ has the intensity with a shape of a central spot surrounded by three light rings, the mode $\mathbf{IG}^{e}_{6,2}$ has the intensity with 4 petals on each of three light rings, the mode $\mathbf{IG}^{e}_{6,4}$ has the intensity with 8 petals on two light rings, and the mode $\mathbf{IG}^{e}_{6,6}$ has the intensity with 12 petals on a single light ring. The intensity pattern of the mode $\mathbf{IG}^{o}_{6,2}$ coincides with that of the mode $\mathbf{IG}^{e}_{6,2}$, rotated by $\pi/4$, intensity pattern of the mode $\mathbf{IG}^{o}_{6,4}$ coincides with that of the mode $\mathbf{IG}^{e}_{6,4}$, rotated by $\pi/8$, and intensity pattern of the mode $\mathbf{IG}^{o}_{6,6}$ coincides with that of the mode $\mathbf{IG}^{e}_{6,6}$, rotated by $\pi/12$.

It is also seen in Fig. 2.5 that when the parameter ε is large and positive, IG modes coincide with corresponding HG modes. Thus, intensity distribution of the mode $\mathbf{IG}^{e}_{6,0}$ has 6 horizontal zero-intensity lines, the mode $\mathbf{IG}^{e}_{6,2}$ has two vertical and four horizontal zero-intensity lines, the mode $\mathbf{IG}^{e}_{6,4}$ has, vice versa, four vertical and two horizontal zero-intensity lines. The intensity of the mode $\mathbf{IG}^{e}_{6,6}$ is the same as that of the mode $\mathbf{IG}^{e}_{6,0}$, but rotated by $\pi/2$. The mode $\mathbf{IG}^{o}_{6,2}$ has the intensity with one vertical and five horizontal zero-intensity lines, the mode $\mathbf{IG}^{o}_{6,4}$ has the intensity with

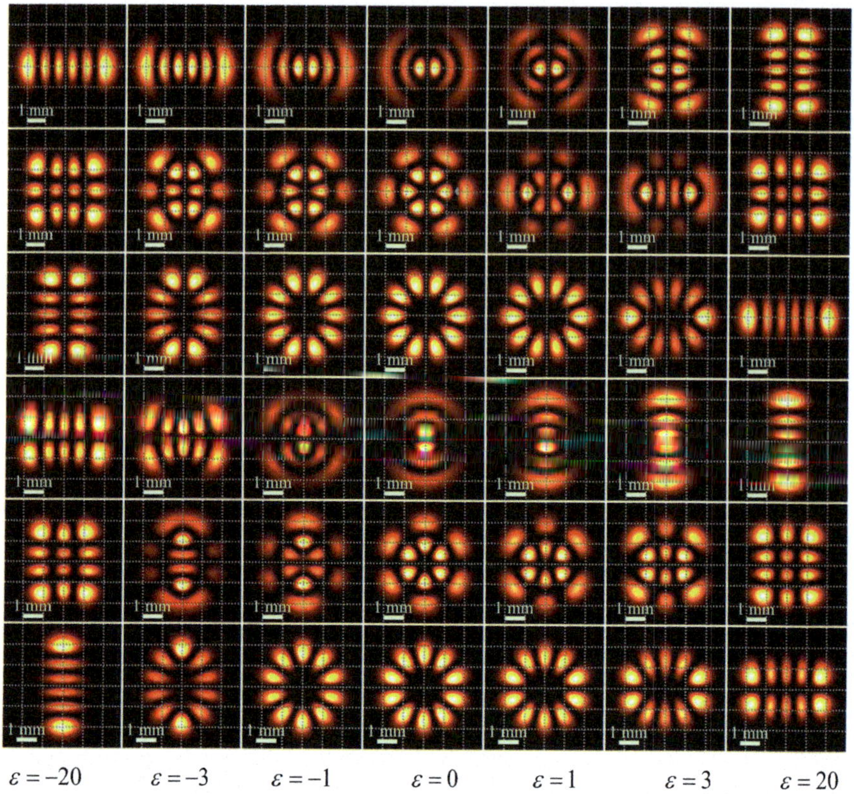

Fig. 2.4 Intensity distributions of the IG modes $\mathbf{IG}^{e}_{5,1}(\mathbf{r}, \varepsilon)$, $\mathbf{IG}^{e}_{5,3}(\mathbf{r}, \varepsilon)$, $\mathbf{IG}^{e}_{5,5}(\mathbf{r}, \varepsilon)$ (rows 1–3), $\mathbf{IG}^{o}_{5,1}(\mathbf{r}, \varepsilon)$, $\mathbf{IG}^{o}_{5,3}(\mathbf{r}, \varepsilon)$, $\mathbf{IG}^{o}_{5,5}(\mathbf{r}, \varepsilon)$ (rows 4–6) for some values of the parameter ε. Other parameters are the following: wavelength $\lambda = 532$ nm, Gaussian beam wais radius $w_0 = 1$ mm

three horizontal and three vertical zero-intensity lines, and, finally, the intensity of the mode $\mathbf{IG}^{o}_{6,6}$ coincides with that of the mode $\mathbf{IG}^{o}_{6,2}$, but rotated by π/2.

In this section, we derived analytical expressions for 24 IG modes with the indices $p = 3, 4, 5, 6$ via 2, 3 or 4 LG modes (more exactly, via their real or imaginary parts) and HG modes (see Eqs. (2.19), (2.25), (2.26), (2.30), (2.34) and Table 2.1). The expansion coefficients of the IG modes by the LG and HG modes are expressed via the ellipticity parameter ε. Using the expressions obtained, one can immediately derive expressions for the IG modes when the parameter ε either equals zero or tends to plus (or minus) infinity. Explicit dependence of the IG modes on the ellipticity parameter allows controlling their intensity by varying this parameter. We have derived the symmetry properties for the even and odd IG modes [21]. In particular, we demonstrated how these modes are interrelated when the ellipticity parameter changes its sign. The obtained representations of the IG modes via the HG and LG modes are not only convenient from the theoretical point of view, since they reveal the properties of these beams without simulation, but also from the practical point

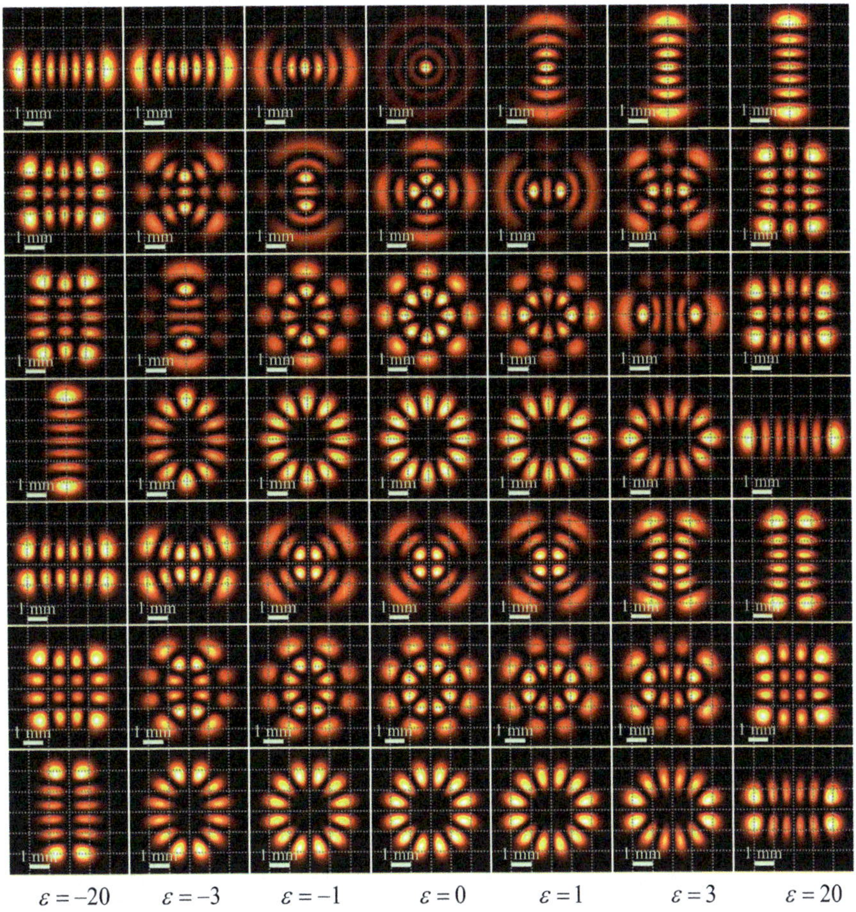

Fig. 2.5 Intensity distributions of the IG modes $\mathbf{IG}^{e}_{6,0}(\mathbf{r}, \varepsilon)$, $\mathbf{IG}^{e}_{6,2}(\mathbf{r}, \varepsilon)$, $\mathbf{IG}^{e}_{6,4}(\mathbf{r}, \varepsilon)$, $\mathbf{IG}^{e}_{6,6}(\mathbf{r}, \varepsilon)$ (rows 1–4), $\mathbf{IG}^{o}_{6,2}(\mathbf{r}, \varepsilon)$, $\mathbf{IG}^{o}_{6,4}(\mathbf{r}, \varepsilon)$, $\mathbf{IG}^{o}_{6,6}(\mathbf{r}, \varepsilon)$ (rows 5–7) for some values of the parameter ε. Other parameters are the following: wavelength $\lambda = 532$ nm, Gaussian beam wais radius $w_0 = 1$ mm

of view, since they allow rather easily programming complex amplitudes of the IG modes for their numerical simulation or generation by using a spatial light modulator. For large-distance atmosphere probing with using the IG beams, high-power laser beams should be generated [22]. The Ince-Gaussian beams can be effectively used for encoding (or protecting) data in wireless information transmission, when the same beam $\mathbf{IG}^{(e,o)}_{p,q}(\mathbf{r}, \varepsilon)$ with the same numbers p, q is transmitted, but the parameter ε is randomly changed.

2.2 Helical Ince-Gaussian Laser Beams as Superposition of Hermite-Gaussian Beams

It has been known [3, 4, 23] that even and odd Ince-Gaussian (IG) modes are real-valued, which means that they are non-vortical beams and carry no orbital angular momentum (OAM). Meanwhile, helical IG (hIG) modes [4, 11], $\mathbf{hIG}_{p,q}(x, y, \varepsilon)$, where p and q are nonnegative integers and ε is a real number (the ellipticity parameter) are complex-valued as on-axis superposition of normalized even and odd IG modes with a phase shift of $\pi/2$. Although the OAM of these modes has been numerically shown to be ε-dependent [11], no relevant analytical relationship was derived. A problem of evaluation of the topological charge (TC) of the hIG modes has not been analyzed also. For example, it is interesting to examine would be whether or not the TC conserves with varying ε. Although an attempt to represent the hIG modes as superposition of Laguerre-Gaussian (LG) modes has been reported [24], an explicit form of the expansion coefficients has not been found. In [21], we derived an explicit form of coefficients of an expansion of IG modes in terms of Hermite-Gaussian (HG) and LG modes at $p \leq 6$.

Both non-helical and helical IG modes have found many uses in optics. For instance, polarization structure of IG beams has been analyzed in [25]. The IG modes can be experimentally realized using a variety of techniques, including a metasurface-aided element [26], a dynamic approach based on a computer-controlled adaptive mirror with a microcell array [27], and spontaneous parametric conversion [9]. The conversion-based technique [9] has been reported to enable generating entangled two-photon hIG modes, which show promise for quantum informatics [11]. Helical IG modes can also be created using a Dove prism and a Mach-Zahnder interferometer [28], or a laser-diode-pumped microchip solid-state laser [29] and other types of lasers [30]. Second-harmonic IG beams can also be generated upon nonlinear conversion of light in a crystal [31]. Classically entangled IG modes were discussed in [8], where Poincaré beams were generated from hIG modes instead of LG modes. The propagation of IG beams through a turbulent atmosphere has also been studied [32]. Alongside quantum informatics and atmospheric sensing, IG beams have paved their way into an area of micromanipulation [33]. The IG beams have been revealed to exhibit a diversity of interesting properties as they propagate in various media, including strongly nonlocal nonlinear media [34] and elliptical core few-mode fibers [35]. The use of IG modes for the effective interatomic interaction has also been reported [36].

For completeness sake, we note that in terms of both their properties and an intensity pattern structure, IG beams are similar to Hermite–Laguerre-Gaussian beams [37–40], as the latter also depend on a real parameter of astigmatism degree and are converted from HG beams into LG beams with parameter variations.

It is interesting enough that although hIG modes have long been around in optics, a number of related theoretical issues has remained unaddressed so far. It has not been clarified yet whether or not the TC of helical IG modes depends on the ellipticity parameter ε. Whether or not, in the general case, the OAM of a hIG mode has any kind of symmetry relative to the parameter ε. Why is the OAM increasing with increasing ε at some values of q while decreasing at other values of q. This work is an attempt to address some of the above-listed questions. We obtain specific relationships for the OAM of some hIG modes, which include the ε-dependence and can be analyzed prior to numerically modeling. We also prove that the OAM carried by an arbitrary hIG mode is an even function of ε and show that the TC at $\varepsilon = 0$ can be different from that at $\varepsilon = \infty$.

2.2.1 Conventional Normalized Ince-Gaussian Modes

Ince-Gaussian (IG) modes [3, 4, 23] present structurally stable solutions of a paraxial Helmholtz equation, $\{\partial_x^2+\partial_y^2+4i\partial_z\}F(\mathbf{r}, z) = 0$, which are separable in the parabolic coordinates (ξ, η) and given by

$$IG_{p,q}(\mathbf{r}, z) = G(\mathbf{r}, z)\left(\frac{1-iz}{1+iz}\right)^{p/2} N_{p,q}(i\xi, \varepsilon)N_{p,q}(\eta, \varepsilon), \tag{2.43}$$

where $\mathbf{r} = (x, y)$ is a 2D vector, $G(\mathbf{r}, z) = \frac{1}{1+iz}\exp\left(-\frac{r^2}{1+iz}\right)$ is the standard Gaussian solution of the paraxial equation and $N_{p,q}(t, \varepsilon)$ is a trigonometric Ince polynomial. Hereinafter, all light fields are analyzed in dimensionless variables, using the following substitutions: $x := x/w_0$, $y := y/w_0$, $z := 2z/(kw_0^2)$, where w_0 is the Gaussian beam waist radius. Besides, we use HG, LG, and IG modes in both normalized (in bold face) and unnormalized notation (in italics). In particular,

$$\mathbf{IG}_{p,q}^{(e,o)}(\mathbf{r}, \varepsilon) = \frac{IG_{p,q}^{(e,o)}(\mathbf{r}, \varepsilon)}{\left\| IG_{p,q}^{(e,o)}(\mathbf{r}, \varepsilon)\right\|}. \tag{2.44}$$

(If the propagation variable, z, is not specified, it is assumed to be zero, i.e. the beam is considered in the waist plane.) The superscript (e or o) denotes the parity of the IG mode, even or odd. The ellipticity parameter ε is real. Its variation from $-\infty$ to $+\infty$ makes it possible to change the IG mode intensity pattern.

We recall the expressions of HG and LG modes in terms of Hermite and Laguerre polynomials, respectively:

$$HG_{n,m}(\mathbf{r}, z) = G(\mathbf{r}, z)\left(\frac{1-iz}{1+iz}\right)^{\frac{n+m}{2}} H_n\left(\frac{x\sqrt{2}}{\sqrt{1+z^2}}\right)H_m\left(\frac{y\sqrt{2}}{\sqrt{1+z^2}}\right),$$

$$LG_{n,\pm m}(\mathbf{r}, z) = G(\mathbf{r}, z)\left(\frac{1-iz}{1+iz}\right)^{n+\frac{m}{2}} L_n^m\left(\frac{2r^2}{1+z^2}\right)r^m e^{\pm im\varphi}, \tag{2.45}$$

$$\mathbf{HG}_{n,m}(\mathbf{r}) = \frac{HG_{n,m}(\mathbf{r})}{\left\| HG_{n,m}(\mathbf{r}) \right\|},$$
$$\left\| HG_{n,m}(\mathbf{r}) \right\| = \sqrt{\frac{\pi}{2} \cdot 2^{n+m} n! m!},$$
$$\mathbf{LG}_{n,\pm m}(\mathbf{r}) = \frac{LG_{n,\pm m}(\mathbf{r})}{\left\| LG_{n,\pm m}(\mathbf{r}) \right\|},$$
$$\left\| LG_{n,\pm m}(\mathbf{r}) \right\| = \sqrt{\frac{\pi}{2} \cdot \frac{(n+m)!}{2^m n!}}. \tag{2.46}$$

It is well known [see Eq. (2.45)] that the HG mode indices n, m determine the number of horizontal and vertical zero-intensity lines, while the radial and azimuthal indices, n, $\pm m$ of the LG modes are for the number of zero intensity rings and the topological charge of the optical vortex at center. For IG modes, the indices p, q are also related to the number of zero-intensity lines—which may be infinite (hyperbolas and straight lines, the index q) or finite (ellipses, the value $(p-q)/2$). A presence of the straight line $y = 0$ among the zero intensity lines clearly indicate that this is an odd IG mode.

In the limiting cases of $\varepsilon = 0$ and $\varepsilon = \pm\infty$, the IG modes are reduced to LG and HG modes. If $\varepsilon = 0$, then

$$\mathbf{IG}^e_{p,q}(\mathbf{r}, 0) = \sqrt{2}\mathrm{Re}\mathbf{LG}_{(p-q)/2,q}(\mathbf{r}),$$
$$\mathbf{IG}^o_{p,q}(\mathbf{r}, 0) = \sqrt{2}\mathrm{Im}\mathbf{LG}_{(p-q)/2,q}(\mathbf{r}), \tag{2.47}$$

except for the case $(p, q) = (2n, 0)$, when $\mathbf{IG}^e_{2n,0}(\mathbf{r}, 0) = \mathbf{LG}_{n,0}(\mathbf{r})$ and there is no an odd IG mode. If $\varepsilon = +\infty$, then

$$\mathbf{IG}^e_{p,q}(\mathbf{r}, +\infty) = (-1)^{(p-q)/2}\mathbf{HG}_{q,p-q}(\mathbf{r}),$$
$$\mathbf{IG}^o_{p,q}(\mathbf{r}, +\infty) = (-1)^{(p-q)/2}\mathbf{HG}_{q-1,p+1-q}(\mathbf{r}). \tag{2.48}$$

If $\varepsilon = -\infty$, then

$$\mathbf{IG}^e_{p,q}(\mathbf{r}, -\infty) = \begin{cases} (-1)^{p/2}\mathbf{HG}_{p-q,q}(\mathbf{r}), & p = 2k, \\ (-1)^{(p-1)/2}\mathbf{HG}_{p+1-q,q-1}(\mathbf{r}), & p = 2k+1, \end{cases}$$
$$\mathbf{IG}^o_{p,q}(\mathbf{r}, -\infty) = \begin{cases} (-1)^{p/2+1}\mathbf{HG}_{p+1-q,q-1}(\mathbf{r}), & p = 2k, \\ (-1)^{(p-1)/2}\mathbf{HG}_{p-q,q}(\mathbf{r}), & p = 2k+1. \end{cases} \tag{2.49}$$

For all other values of ε, the IG modes are no longer reduced to a single LG or HG mode but to a superposition of such modes. If we choose the basis of LG modes, then $\mathbf{IG}^e_{p,q}(\mathbf{r}, \varepsilon)$ and $\mathbf{IG}^o_{p,q}(\mathbf{r}, \varepsilon)$ modes are, respectively, expanded in terms of real and imaginary parts of the modes $\mathbf{LG}_{k,p-2k}(\mathbf{r})$, where the summation index k runs through values from 0 to $[p/2]$. When the basis of HG modes is selected, $\mathbf{IG}^e_{p,q}(\mathbf{r}, \varepsilon)$ and $\mathbf{IG}^o_{p,q}(\mathbf{r}, \varepsilon)$ are expanded into series of modes $\mathbf{HG}_{p-k,k}(\mathbf{r})$, where $0 \le k \le p$;

however, for even IG modes, the summation index k runs through even values only, while for odd IG modes it runs through odd values.

For $p \leq 6$, we have found such expansions in our paper [21]. In particular, for even and odd IG modes with the indices $(p, q) = \{(2, 2),\ (3, 1),\ (3, 3)\}$, the formulae are as follows:

$$\mathbf{IG}^e_{2,2}(\mathbf{r}, \varepsilon) = \frac{a \cdot \mathbf{HG}_{2,0}(\mathbf{r}) - \mathbf{HG}_{0,2}(\mathbf{r})}{\sqrt{1+a^2}},$$
$$\mathbf{IG}^o_{2,2}(\mathbf{r}, \varepsilon) = \mathbf{HG}_{1,1}(\mathbf{r}). \tag{2.50}$$

$$\mathbf{IG}^e_{3,1}(\mathbf{r}, \varepsilon) = -\frac{\sqrt{3} \cdot \mathbf{HG}_{3,0}(\mathbf{r}) + b \cdot \mathbf{HG}_{1,2}(\mathbf{r})}{\sqrt{3+b^2}}$$
$$= \frac{\sqrt{3}(b-1)\mathrm{Re}\mathbf{LG}_{0,3}(\mathbf{r}) + (b+3)\mathrm{Re}\mathbf{LG}_{1,1}(\mathbf{r})}{\sqrt{2} \cdot \sqrt{3+b^2}}, \tag{2.51}$$

$$\mathbf{IG}^o_{3,1}(\mathbf{r}, \varepsilon) = -\frac{\sqrt{3} \cdot \mathbf{HG}_{2,1}(\mathbf{r}) + c \cdot \mathbf{HG}_{0,3}(\mathbf{r})}{\sqrt{3+c^2}}$$
$$= \frac{(c-3)\mathrm{Im}\mathbf{LG}_{0,3}(\mathbf{r}) + \sqrt{3}(c+1)\mathrm{Im}\mathbf{LG}_{1,1}(\mathbf{r})}{\sqrt{2} \cdot \sqrt{3+c^2}}, \tag{2.52}$$

$$\mathbf{IG}^e_{3,3}(\mathbf{r}, \varepsilon) = +\frac{b \cdot \mathbf{HG}_{3,0}(\mathbf{r}) - \sqrt{3} \cdot \mathbf{HG}_{1,2}(\mathbf{r})}{\sqrt{3+b^2}}$$
$$= \frac{(b+3)\mathrm{Re}\mathbf{LG}_{0,3}(\mathbf{r}) - \sqrt{3}(b-1)\mathrm{Re}\mathbf{LG}_{1,1}(\mathbf{r})}{\sqrt{2} \cdot \sqrt{3+b^2}}, \tag{2.53}$$

$$\mathbf{IG}^o_{3,3}(\mathbf{r}, \varepsilon) = +\frac{c \cdot \mathbf{HG}_{2,1}(\mathbf{r}) - \sqrt{3} \cdot \mathbf{HG}_{0,3}(\mathbf{r})}{\sqrt{3+c^2}}$$
$$= \frac{\sqrt{3}(c+1)\mathrm{Im}\mathbf{LG}_{0,3}(\mathbf{r}) - (c-3)\mathrm{Im}\mathbf{LG}_{1,1}(\mathbf{r})}{\sqrt{2} \cdot \sqrt{3+c^2}}, \tag{2.54}$$

where

$$a = a(\varepsilon) = \varepsilon + \sqrt{\varepsilon^2 + 1},$$
$$b = b(\varepsilon) = \varepsilon - 1 + \sqrt{(\varepsilon - 1)^2 + 3},$$
$$c = c(\varepsilon) = \varepsilon + 1 + \sqrt{(\varepsilon + 1)^2 + 3}. \tag{2.55}$$

Since $a(\varepsilon)a(-\varepsilon) = 1$ and $b(\varepsilon)c(-\varepsilon) = 3$, we can easily check the symmetry properties of IG modes (2.50)–(2.54), when the sign of the ellipticity parameter is reversed:

$$\mathbf{IG}^e_{2n,2m}(x, y, -\varepsilon) = (-1)^m \mathbf{IG}^e_{2n,2m}(y, x, \varepsilon),$$

$$\mathbf{IG}^{o}_{2n,2m}(x, y, -\varepsilon) = (-1)^{m+1}\mathbf{IG}^{o}_{2n,2m}(y, x, \varepsilon),$$
$$\mathbf{IG}^{o}_{2n+1,2m+1}(x, y, -\varepsilon) = (-1)^{m}\mathbf{IG}^{e}_{2n+1,2m+1}(y, x, \varepsilon). \quad (2.56)$$

It can be proven, that this property is valid in general case of indices. The IG modes are real-valued and have no orbital angular momentum (OAM). In the next section, using the expansions (2.50)–(2.54) and similar others with different values of (p, q), we derive expressions of helical IG (hIG) modes.

2.2.2 *Expansion of Helical IG Modes into HG Modes*

Following Refs. [4, 11] we construct hIG modes as superposition of normalized even and odd IG modes shifted in phase by $\pi/2$.

$$\mathbf{hIG}^{\pm}_{p,q}(\mathbf{r}, \varepsilon) = \frac{\mathbf{IG}^{e}_{p,q}(\mathbf{r}, \varepsilon) \pm i\mathbf{IG}^{o}_{p,q}(\mathbf{r}, \varepsilon)}{\sqrt{2}}. \quad (2.57)$$

Without lost of generality, in (2.57) it would suffice to analyze just the sum of the hIG modes. At $(p, q) = \{(2, 2), (3, 1), (3, 3)\}$, in view of Eqs. (2.50)–(2.54), the hIG modes take the form:

$$\begin{aligned}\mathbf{hIG}^{+}_{2,2}(\mathbf{r}, \varepsilon) &= \frac{\mathbf{IG}^{e}_{2,2}(\mathbf{r}, \varepsilon) + i\mathbf{IG}^{o}_{2,2}(\mathbf{r}, \varepsilon)}{\sqrt{2}}\\ &= \frac{1}{\sqrt{2}}\left(\frac{a \cdot \mathbf{HG}_{2,0}(\mathbf{r}) - \mathbf{HG}_{1,2}(\mathbf{r})}{\sqrt{1+a^2}} + i\mathbf{HG}_{1,1}(\mathbf{r})\right),\end{aligned} \quad (2.58)$$

$$\begin{aligned}\mathbf{hIG}^{+}_{3,1}(\mathbf{r}, \varepsilon) &= \frac{\mathbf{IG}^{e}_{3,1}(\mathbf{r}, \varepsilon) + i\mathbf{IG}^{o}_{3,1}(\mathbf{r}, \varepsilon)}{\sqrt{2}}\\ &= -\frac{1}{\sqrt{2}}\left(\frac{\sqrt{3} \cdot \mathbf{HG}_{3,0}(\mathbf{r}) + b \cdot \mathbf{HG}_{1,2}(\mathbf{r})}{\sqrt{3+b^2}}\right.\\ &\left.\quad + i\frac{\sqrt{3} \cdot \mathbf{HG}_{2,1}(\mathbf{r}) + c \cdot \mathbf{HG}_{0,3}(\mathbf{r})}{\sqrt{3+c^2}}\right),\end{aligned} \quad (2.59)$$

$$\begin{aligned}\mathbf{hIG}^{+}_{3,3}(\mathbf{r}, \varepsilon) &= \frac{\mathbf{IG}^{e}_{3,3}(\mathbf{r}, \varepsilon) + i\mathbf{IG}^{o}_{3,3}(\mathbf{r}, \varepsilon)}{\sqrt{2}}\\ &= \frac{1}{\sqrt{2}}\left(\frac{b \cdot \mathbf{HG}_{3,0}(\mathbf{r}) - \sqrt{3} \cdot \mathbf{HG}_{1,2}(\mathbf{r})}{\sqrt{3+b^2}}\right.\\ &\left.\quad + i\frac{c\mathbf{HG}_{2,1}(\mathbf{r}) - \sqrt{3} \cdot \mathbf{HG}_{0,3}(\mathbf{r})}{\sqrt{3+c^2}}\right).\end{aligned} \quad (2.60)$$

With the beams under study being structurally stable modes, their transverse intensity pattern conserves upon free-space propagation, changing only in scale.

2.2.3 Orbital Angular Momentum of hIG Modes

Expanding the hIG modes into a series in terms of HG modes enables their orbital angular momentum (OAM) to be derived at any value of the parameter ε. It has been known [41] that for any structurally stable field defined in the waist plane by a series in terms of HG modes, $E(\mathbf{r}) = \sum_{0\le m\le N} C_m HG_{N-m,m}(\mathbf{r})$, its OAM can be found from

$$\mathrm{OAM}[E(\mathbf{r})] = \frac{2\sum_{1\le m\le N} \mathrm{Im}(C_{m-1}^* C_m)(N+1-m)!m!}{\sum_{0\le m\le N} |C_m|^2 (N-m)!m!}. \tag{2.61}$$

For a normalized version of the field, $\mathbf{E}(\mathbf{r}) = \sum_{0\le m\le N} \tilde{C}_m \mathbf{HG}_{N-m,m}(\mathbf{r})$, Eq. (2.61) is rearranged to

$$\mathrm{OAM}[\mathbf{E}(\mathbf{r})] = 2\sum_{1\le m\le N} \mathrm{Im}(\tilde{C}_{m-1}^* \tilde{C}_m)\sqrt{(N+1-m)m}. \tag{2.62}$$

Applying Eq. (2.62) to hIG modes in (2.58)–(2.60) yields:

$$\begin{aligned}
\mathrm{OAM}\big[\mathbf{hIG}_{2,2}^{+}(\mathbf{r},\varepsilon)\big] &= \frac{(1+a)\sqrt{2}}{\sqrt{1+a^2}},\\
\mathrm{OAM}\big[\mathbf{hIG}_{3,1}^{+}(\mathbf{r},\varepsilon)\big] &= \frac{(3-2b+bc)\sqrt{3}}{\sqrt{(3+b^2)(3+c^2)}},\\
\mathrm{OAM}\big[\mathbf{hIG}_{3,3}^{+}(\mathbf{r},\varepsilon)\big] &= \frac{(3+2c+bc)\sqrt{3}}{\sqrt{(3+b^2)(3+c^2)}},
\end{aligned} \tag{2.63}$$

where the parameters *a*, *b*, *c* are defined in (2.55).

At the limiting cases, $\varepsilon = 0$ and $\varepsilon = \pm\infty$, the hIG modes are, respectively, described by a single standard LG mode and a complex sum of two HG modes, owing to Eqs. (2.47)–(2.49):

$$\begin{aligned}
\mathbf{hIG}_{p,q}^{+}(\mathbf{r},0) &= \mathbf{LG}_{(p-q)/2,q}(\mathbf{r}),\\
\mathbf{hIG}_{p,q}^{+}(\mathbf{r},+\infty) &= \frac{(-1)^{(p-q)/2}}{\sqrt{2}}\big\{\mathbf{HG}_{q,p-q}(\mathbf{r}) + i\mathbf{HG}_{q-1,p+1-q}(\mathbf{r})\big\},\\
\mathbf{hIG}_{p,q}^{+}(\mathbf{r},-\infty) &= \frac{i^{p-1}}{\sqrt{2}}\big\{\mathbf{HG}_{p+1-q,q-1}(\mathbf{r}) + i\mathbf{HG}_{p-q,q}(\mathbf{r})\big\}.
\end{aligned} \tag{2.64}$$

At $\varepsilon = 0$, the OAM of an hIG mode equals the topological charge of an LG mode: OAM $= q$. The cases $\varepsilon = \pm\infty$ are simple also, since the series in (2.62) reduces to a single term, namely, $m = p + 1 - q$ at $\varepsilon = +\infty$ and $m = q$ at $\varepsilon = -\infty$. So, for the beams in (2.64) we find that

$$\begin{aligned} \text{OAM}\left[\mathbf{hIG}^{+}_{p,q}(\mathbf{r}, 0)\right] &= q, \\ \text{OAM}\left[\mathbf{hIG}^{+}_{p,q}(\mathbf{r}, \pm\infty)\right] &= \sqrt{q(p+1-q)}. \end{aligned} \tag{2.65}$$

In particular, Eq. (2.65) suggests that at $p = 7$ and $q = 1, 3, 5, 7$, an hIG mode, respectively, carries an OAM of 1, 3, 5, 7 at $\varepsilon = 0$, and $\sqrt{7}, \sqrt{15}, \sqrt{15}, \sqrt{7}$ at $\varepsilon = \pm\infty$. Hence, for the hIG modes, a plot for the OAM as a function of ε drawn at $q = 7$ should have two intersections (at $\varepsilon > 0$ and $\varepsilon < 0$) with similar plots drawn at $q = 3$ and 5. As result, Eq. (2.65) provide a theoretical background and generalization of the numerical result of Ref. [11].

It can be checked easily that Eq. (2.63) are compatible with the general formulae (2.65) by substituting $a(0) = 1$, $b(0) = 1$, and $c(0) = 3$ into (2.63) and using asymptotics $a(\varepsilon) \sim 2\varepsilon$, $b(\varepsilon) \sim 2\varepsilon$, $c(\varepsilon) \sim 2\varepsilon$ at $\varepsilon \to +\infty$.

Equations (2.63)–(2.65) are one of the main results of this section, since they provide analytical ties between the OAM and the parameter ε of hIG modes at small values of the indices (p, q). It should be noted that using a technique described in Ref. [21] analytical relationships for the OAM of the hIG modes at higher values of p, $p > 3$, can also be deduced, but when p grows, they become more and more cumbersome. Below, relationships for the OAM of hIG modes at $p = 4$ and 5 are given without a proof.

2.2.4 Topological Charge of hIG Modes

Using a decomposition of hIG modes into LG modes in (2.51)–(2.54) and others bases [21], we can determine a topological charge (TC) of the hIG modes at $\varepsilon = 0$, TC $= q$. For the hIG modes in (2.58)–(2.60), this result can be verified directly by putting $\varepsilon = 0$ and changing to the LG modes in (2.51)–(2.54).

It would be interesting to reveal whether or not the TC of hIG modes varies with varying parameter ε. From (2.64) it is seen that at $\varepsilon = +\infty$, any hIG mode can be expressed as superposition of two HG modes, meaning that, in view of (2.45), it has zeroes at zeroes of a polynomial of two variables:

$$H_q(X)H_{p-q}(Y) + i \cdot H_{q-1}(X)H_{p+1-q}(Y), \tag{2.66}$$

where we use a substitution $X = x\sqrt{2}$ and $Y = y\sqrt{2}$ for brevity. The sum in (2.66) vanishes in two cases:

1. $H_q(X) = H_{p+1-q}(Y) = 0$,
2. $H_{q-1}(X) = H_{p-q}(Y) = 0$.

Since any Hermite polynomial $H_n(t)$ has n simple (nondegenerate) real zeroes, which we denote as $t_{n,k}$, where $k = 1, 2, \ldots, n$, then for the case (1), there is an array of $q(p+1-q)$ isolated zeroes $(X_0, Y_0) = (t_{q,\nu}, t_{p+1-q,\mu})$, where $1 \leq \nu \leq q$ and $1 \leq \mu \leq p+1-q$, while for the case (2), there is an array of $(q-1)(p-q)$ isolated zeroes $(X_0, Y_0) = (t_{q-1,\nu}, t_{p-q,\mu})$, where $1 \leq \nu \leq q-1$ and $1 \leq \mu \leq p-q$. It can be proven that in the case (1), all zeroes are of type $(X+iY)$, i.e. have TC $= +1$, whereas in the case (2) they are all of type $(X - iY)$, i.e. TC $= -1$. Actually, an analysis of relationships $H'_n(t) = 2nH_{n-1}(t)$ and $H_{n+1}(t) = 2tH_n(t) - 2nH_{n-1}(t)$ shows that at fixed values of n and k, the expressions $H_{n-1}(t_{n,k})$, $H_{n+1}(t_{n,k})$, and $H'_n(t_{n,k})$ are non-zero. Moreover, $H_{n-1}(t_{n,k})$ and $H'_n(t_{n,k})$ have the same sign, but $H_{n+1}(t_{n,k})$, $H'_n(t_{n,k})$ have opposite signs. Then expanding the expression (2.66) into a Taylor series up to linear terms in the vicinity of each isolated zero (X_0, Y_0), we find that for the case (1) coefficients by $(X - X_0)$ and $i(Y - Y_0)$ are of the same sign and for the case (2) coefficients by $(Y - Y_0)$ and $i(X - X_0)$ are of the same sign. So, we can deduce that at $\varepsilon = +\infty$, the mode $\mathbf{hIG}^+_{p,q}$ has TC $= q(p+1-q) - (q-1)(p-q) = p$. At $\varepsilon = -\infty$, the TC is derived in a similar manner and equals TC $= p$.

Thus, we obtain that the TC of the mode $\mathbf{hIG}^+_{p,q}$ is ε-independent at $q = p$. However, if $q < p$, then the TC increases with increasing $|\varepsilon|$ from q at $\varepsilon = 0$ to p at $\varepsilon = \pm\infty$. It is not surprising that the TC increases with increasing $|\varepsilon|$. As an example, the series expansion of $\mathbf{hIG}^+_{3,1}$ in terms of LG modes

$$\mathbf{hIG}^+_{3,1}(\mathbf{r}, \varepsilon) = \frac{\sqrt{3}(b-1)\mathrm{Re}\mathbf{LG}_{0,3}(\mathbf{r}) + (b+3)\mathrm{Re}\mathbf{LG}_{1,1}(\mathbf{r})}{2\sqrt{3+b^2}} + i \cdot \frac{(c-3)\mathrm{Im}\mathbf{LG}_{0,3}(\mathbf{r}) + \sqrt{3}(c+1)\mathrm{Im}\mathbf{LG}_{1,1}(\mathbf{r})}{2\sqrt{3+c^2}}, \tag{2.67}$$

shows that when the polar radius r and parameter ε both tend to infinity, Eq. (2.67) is rearranged to an expression which depends only on the azimuthal angle φ (neglecting a common Gaussian factor):

$$\mathbf{hIG}^+_{3,1}(\mathbf{r}, +\infty)\Big|_{r\to\infty} \sim (\sqrt{3}\cos 3\varphi + i \sin 3\varphi) + (\cos\varphi + i\sqrt{3}\sin\varphi). \tag{2.68}$$

According to Ref. [42], the TC of the sum of two optical vortices with different coefficients equals that of the constituent vortex with the larger coefficient, and in the case of equal coefficients, equals the larger TC of the two constituent vortices. Thus, the mode in (2.68) has the TC $= 3$, notwithstanding that at $\varepsilon = 0$ the TC $= 1$.

2.2.5 Parity of the OAM Function in Terms of the Parameter ε

Since both even and odd IG components in the definition (2.57) of the hIG mode are real-valued, the reverse of the sign between the components leads to a complex conjugated mode, also resulting in the change of the OAM:

$$\mathrm{OAM}\Big[\mathbf{hIG}^{-}_{p,q}(\mathbf{r}, \varepsilon)\Big] = -\mathrm{OAM}\Big[\mathbf{hIG}^{+}_{p,q}(\mathbf{r}, \varepsilon)\Big]. \tag{2.69}$$

On the other hand, it is seen from (2.56) that the reverse of the parameter ε sign leads to a conjugated hIG mode, with x and y being interchanged:

$$\mathbf{hIG}^{+}_{p,q}(x, y, -\varepsilon) = (-1)^m \begin{cases} \mathbf{hIG}^{+}_{p,q}(y, x, \varepsilon), & (p, q) = (2n, 2m) \\ \pm i \cdot \mathbf{hIG}^{+}_{p,q}(y, x, \varepsilon), & (p, q) = (2n+1, 2m+1). \end{cases} \tag{2.70}$$

Interchanging the variables x and y leads to the change of sign of the rotation operator, $x\partial_y - y\partial_x$, whereas the replacement of ε by $-\varepsilon$ leads to a conjugation of the hIG mode. As result, when ε is replaced by $-\varepsilon$, the OAM of the original hIG mode remains unchanged. So, the OAM of any hIG mode is an even function of the parameter ε:

$$\mathrm{OAM}\Big[\mathbf{hIG}^{+}_{p,q}(\mathbf{r}, -\varepsilon)\Big] = \mathrm{OAM}\Big[\mathbf{hIG}^{+}_{p,q}(\mathbf{r}, \varepsilon)\Big]. \tag{2.71}$$

It would be interesting to figure out if the OAM of the hIG mode achieves its maximum at $\varepsilon = 0$. The question is not unfounded as at $\varepsilon = 0$, any hIG mode can be reduced to a single LG mode, Eq. (2.64). We note that at $\varepsilon > 0$, an expansion of the hIG mode always contains several LG modes with different topological charges, which may be expected to result in a decrease in the OAM due to a redistribution between several constituent modes. From (2.65) it follows that at $q < (p+1)/2$, the OAM is higher at $\varepsilon = \pm\infty$ than at $\varepsilon = 0$, and, vice versa, at $q > (p+1)/2$, the OAM is lower at infinity than at $\varepsilon = 0$.

2.2.6 Relationships for the OAM of hIG Modes at P = 4 and 5

Here, we present relationships for the OAM of hIG modes at $p = 4$ and 5. In both cases, we need to find characteristic values as roots of certain cubic equations (see below). Since the expressions of the roots themselves and of the corresponding even and odd IG modes are quite cumbersome (all of them can be found in our work [21]), we do not give the formulae below, writing down only relationships for the OAM.

Let $p = 4$. Then, $q = 2m = 0, 2, 4$. At $q = 0$, the mode $\mathbf{hIG}^{+}_{4,0}(\mathbf{r}, \varepsilon)$ is real-valued, meaning that the OAM = 0. At $q = 2$ and $q = 4$, we obtain:

$$\begin{aligned}\text{OAM}\big[\mathbf{hIG}^{+}_{4,2}(\mathbf{r}, \varepsilon)\big] &= \frac{4\varepsilon(d-1) - (\lambda^e_1 - 16)(d+1)}{\sqrt{D(\lambda^e_1, \varepsilon)} \cdot \sqrt{1+d^2}} \cdot \varepsilon\sqrt{6},\\ \text{OAM}\big[\mathbf{hIG}^{+}_{4,4}(\mathbf{r}, \varepsilon)\big] &= \frac{4\varepsilon(d+1) + (\lambda^e_2 - 16)(d-1)}{\sqrt{D(\lambda^e_2, \varepsilon)} \cdot \sqrt{1+d^2}} \cdot \varepsilon\sqrt{6}.\end{aligned} \tag{2.72}$$

Here, $d = d(\varepsilon) = (\varepsilon + \sqrt{\varepsilon^2 + 9})\big/3$,

$$D(\lambda, \varepsilon) = (16 + 5\varepsilon^2)\lambda^2 - 8(40 + 23\varepsilon^2)\lambda + 16(64 + 100\varepsilon^2 + \varepsilon^4)$$

and $\lambda^e_m(\varepsilon)(m = 0, 1, 2)$ are roots of a characteristic equation for even IG modes:

$$\lambda^3 - 20\lambda^2 + (64 - 16\varepsilon^2)\lambda + 192\varepsilon^2 = 0, \tag{2.73}$$

enumerated in ascending order, $\lambda^e_0 < \lambda^e_1 < \lambda^e_2$. In particular, $\lambda^e_0(0) = 0$, $\lambda^e_1(0) = 4$, $\lambda^e_2(0) = 16$.

Let $p = 5$. Then, $q = 2m + 1 = 1, 3, 5$ and relationships for the OAM get even more cumbersome:

$$\text{OAM}\big[\mathbf{hIG}^{+}_{5,2m+1}(\mathbf{r}, \varepsilon)\big] = \frac{Q}{4\sqrt{D(\lambda^e_m(\varepsilon), \varepsilon)} \cdot \sqrt{D(\lambda^o_m(\varepsilon), -\varepsilon)}}. \tag{2.74}$$

Here,

$$\begin{aligned} Q &= \big\{(\lambda^e_m - 9)(\lambda^e_m - 25) - 5\varepsilon^2\big\}\big\{(\lambda^o_m - 9)(\lambda^o_m - 25) - 5\varepsilon^2\big\}\\ &\quad + 24\varepsilon^2(\lambda^e_m - 25)(\lambda^o_m - 25) + 200\varepsilon^4,\\ D(\lambda, \varepsilon) &= (48 - 24\varepsilon + 5\varepsilon^2)\lambda^2 - 2(816 - 408\varepsilon + 125\varepsilon^2 - 3\varepsilon^3)\lambda\\ &\quad + 5(2160 - 1080\varepsilon + 577\varepsilon^2 - 6\varepsilon^3 + \varepsilon^4),\end{aligned}$$

$\lambda^e_m(\varepsilon)(m = 0, 1, 2)$ are roots of the characteristic equation for even IG modes:

$$\begin{aligned}&\lambda^3 - (35 + 3\varepsilon)\lambda^2 + (259 + 102\varepsilon - 13\varepsilon^2)\lambda\\ &\quad - (225 + 675\varepsilon - 205\varepsilon^2 - 15\varepsilon^3) = 0,\end{aligned} \tag{2.75}$$

and $\lambda^o_m(\varepsilon)(m = 0, 1, 2)$ are roots of the characteristic equation for odd IG modes, which is obtained from (A4) by substituting $\varepsilon \to -\varepsilon$. As usual, the roots are numbered in ascending order. In particular, $\lambda^{(e,o)}_0(0) = 1$, $\lambda^{(e,o)}_1(0) = 9$, $\lambda^{(e,o)}_2(0) = 25$.

We note that for the limiting cases, $\varepsilon = 0$ and $\varepsilon = \pm\infty$, the OAM evaluation from (2.72) and (2.74) requires to resolve the indeterminacy $0\big/0$ or ∞/∞. For instance, at $(p, q) = (4, 2)$, if $\varepsilon \sim 0$, then, $d \sim 1 + \varepsilon/3$, $\lambda_1 \sim 4 + 8\varepsilon^2/3 - 56\varepsilon^4/27$, $D(\lambda_1, \varepsilon) \sim 432\varepsilon^2 - 2288\varepsilon^4/9$. Hence, $\mathrm{OAM}[\mathbf{hIG}^{+}_{4,2}(\mathbf{r}, \varepsilon)] \sim 2 + 221\varepsilon^2/972$. In particular, at $\varepsilon = 0$, we obtain a well-known result, $\mathrm{OAM}[\mathbf{hIG}^{+}_{4,2}(\mathbf{r}, 0)] = 2$, since $\mathbf{hIG}^{+}_{4,2}(\mathbf{r}, 0) = \mathbf{LG}_{1,2}(\mathbf{r})$.

2.2.7 Numerical Modeling

Figure 2.6 depicts intensity patterns for the hIG modes numerically simulated using Eq. (2.57) for the indices $p = 7$ and $q = 1, 3, 5, 7$. The intensity patterns are seen to be symmetrical relative to the $y - x$ line for the cases of ε and $-\varepsilon$, which follows from (2.56).

Figure 2.7 depicts patterns of intensity (a–d) and phase (e–h) for modes $\mathbf{hIG}^{+}_{7,q}(\mathbf{r}, +\infty)$ at $q = 1, 3, 5, 7$. The numerical simulation is conducted using (2.64). The waist radius is 1 mm and the frame size is $-R \le x, y \le R$, where $R = 10$ mm. The TC carried by the modes in Fig. 2.7 is calculated using the formula by Berry [43]:

$$\mathrm{TC}[E(r, \varphi)] = \frac{1}{2\pi} \lim_{r \to \infty} \mathrm{Im} \int_0^{2\pi} \frac{\partial_\varphi E(r, \varphi)}{E(r, \varphi)} d\varphi, \tag{2.76}$$

where $E(r, \varphi)$ is the complex amplitude of the hIG mode in the polar coordinates in the waist plane. In our numerical simulation, the TC is calculated on a circle of a radius equal to half the frame size in Fig. 2.7. At different values of the index q, the numerically simulated TC is found to be TC = 6.9966237 ($q = 1$), TC = 6.9973487 ($q = 3$), TC = 6.9973487 ($q = 5$), TC = 6.9966237 ($q = 7$). Hence, for all q analyzed, the TC is seen to be the same and equal to $p = 7$, which is in compliance with the theoretical prediction. Phase distribution patterns in Fig. 2.7 demonstrate

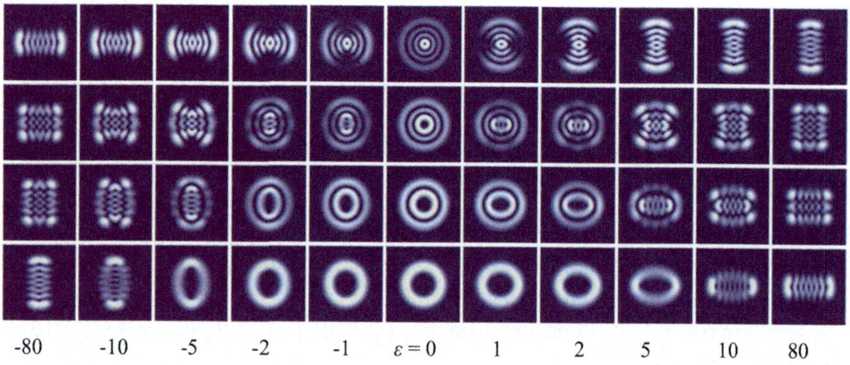

Fig. 2.6 Intensity patterns at the waist of hIG modes (2.57) at $p = 7$ and $q = 1, 3, 5, 7$ (top to bottom) and some values of ε

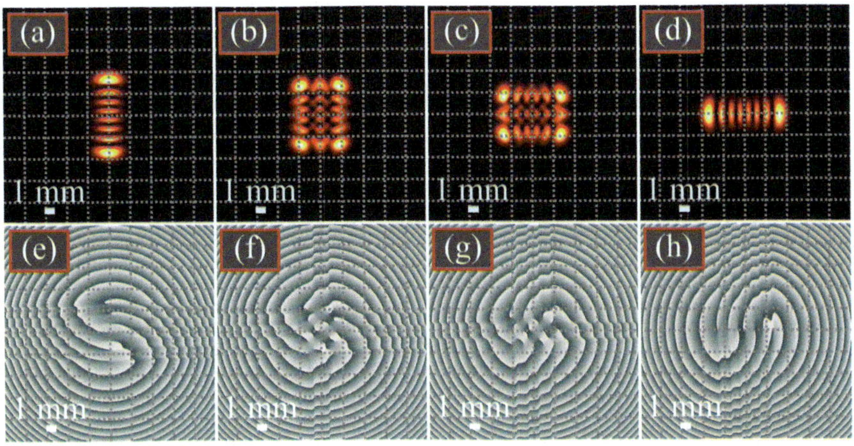

Fig. 2.7 Patterns of intensity (**a**–**d**) and phase (**e**–**h**) for the hIG modes in the original plane at $\varepsilon = +\infty$ for $p = 7$ and q: 1 (**a**, **e**), 3 (**d**, **f**), 5 (**c**, **g**), and 7 (**d**, **h**)

seven lines of phase jump by 2π, with their initial points being located on the vertical axis ($q = 1$), the horizontal axis ($q = 7$), two vertical ($q = 3$) and two horizontal ($q = 5$) lines.

Figures 2.8, 2.9, and 2.10 depict patterns of intensity and phase of the original hIG mode for $(p, q) = (2, 2)$, $(3, 1)$, and $(3, 3)$ at different ε. The Gaussian beam waist radius is $w_0 = 1$ mm and the frame size is $-R \leq x, y \leq R = 3$ mm.

From Figs. 2.8, 2.9, and 2.10, the theoretically predicted and numerically calculated values of the OAM are seen to coincide up to 3 decimal places. Phase patterns suggest that at different values of $q < p$ (Fig. 2.9), the TC does not conserve, increasing from $q = 1$ to $p = 3$ as the parameter ε is increasing from 0 to +20 or decreasing from 0 to −20.

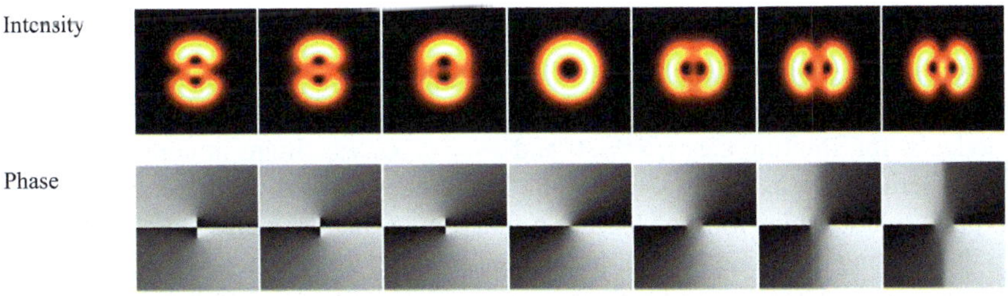

ε	−20	−3	−1	0	1	3	20
OAM,theor.	1.449095	1.622484	1.847759	2.000000	1.847759	1.622484	1.449095
OAM, num	1.449085	1.622474	1.847747	1.999962	1.847747	1.622474	1.449085

Fig. 2.8 Patterns of intensity and phase for the hIG mode $(p, q) = (2, 2)$

Intensity

Phase

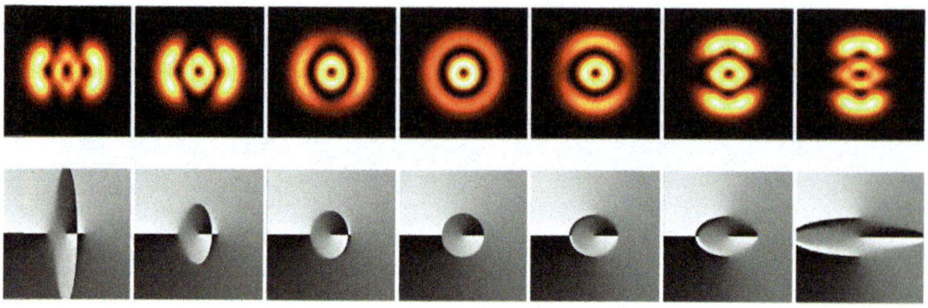

ε	−20	−3	−1	0	1	3	20
OAM,theor.	1.649851	1.331704	1.081395	1.000000	1.081395	1.331704	1.649851
OAM, num.	1.649836	1.331690	1.081385	0.999992	1.081385	1.331690	1.649836

Fig. 2.9 Patterns of intensity and phase for the hIG mode $(p, q) = (3, 1)$

Intensity

Phase

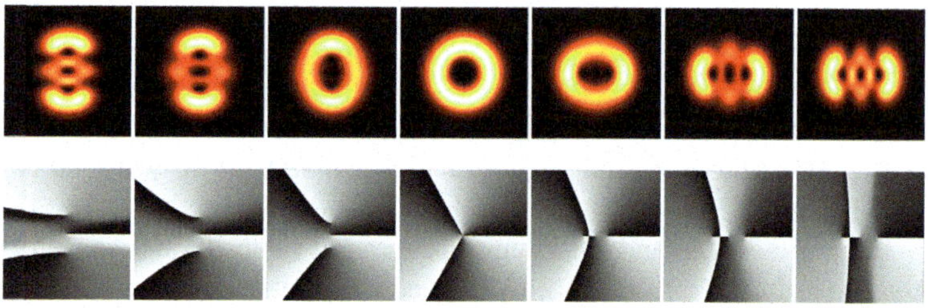

ε	−20	−3	−1	0	1	3	20
OAM, theor.	1.822836	2.396077	2.900545	3.000000	2.900545	2.396077	1.822836
OAM, num.	1.822821	2.396060	2.900522	2.999974	2.900522	2.396060	1.822821

Fig. 2.10 Patterns of intensity and phase for the hIG mode $(p, q) = (3, 3)$

Summing up, the following results have been obtained. Explicit analytical formulae for complex amplitudes of hIG modes at $p = 2, 3$, Eqs. (2.58)–(2.60), have been derived. Using those equations, one can easily deduce expressions for hIG modes at the limiting cases, $\varepsilon = 0$ and $\varepsilon = \pm\infty$, which are in compliance with the general relationships (2.64). For the hIG modes studied, we have also deduced explicit relationships for the OAM as a function of the parameter ε, Eq. (2.63), and in the limiting cases, Eq. (2.60). In the general case, the ε-dependence of the OAM has been shown to be an even function, Eq. (2.71). It has been shown that at $\varepsilon = 0$, the TC of hIG modes equals that of a LG mode, i.e. the index q, being equal to the index p at $\varepsilon = +\infty$. Thus, the TC of hIG modes is not conserved with varying parameter ε.

References

1. F.M. Arscott, *Periodic Differential Equations* (Pergamon, Oxford, 1964)
2. F.M. Arscott, The Whittaker-Hill equation and the wave equation in paraboloidal coordinates. Proc. R. Soc. Edinburgh Sect. A **67**, 265–276 (1967)
3. W. Miller, *Symmetry and separation of variables* (Addison-Wesley Pub. Comp, London, 1977)
4. M.A. Bandres, J.C. Gutiérrez-Vega, Ince-Gaussian beams. Opt. Lett. **29**, 144–146 (2004)
5. M.A. Bandres, J.C. Gutiérrez-Vega, Ince-Gaussian modes of the paraxial wave equation and stable resonators. J. Opt. Soc. Am. A **21**, 873–880 (2004)
6. M.A. Bandres, Elegant Ince-Gaussian beams. Opt. Lett. **29**, 1724–1726 (2004)
7. S.K. Singh, H. Haginaka, B.J. Jackin, K. Kinashi, N. Tsutsumi, W. Sakai, Generation of Ince-Gaussian beams using azocarbazole polymer CGH. J. Imag. **8**, 144 (2022)
8. Y. Li, X.B. Hu, B. Perez-Garcia, B. Zhao, W. Gao, Z.H. Zhu, C. Rosales-Guzmán, Classically entangled Ince-Gaussian modes. Appl. Phys. Lett. **116**, 221105 (2020)
9. B. Baghdasaryan, S. Fritzsche, Enhanced entanglement from Ince-Gaussian pump beams in spontaneous parametric down-conversion. Phys. Rev. A **102**, 052412 (2020)
10. M. Krenn, R. Fickler, M. Huber, R. Lapkiewicz, W. Plick, S. Ramelow, A. Zeilinger, Entangled singularity patterns of photons in Ince-Gauss modes. Phys. Rev. A **87**, 012326 (2013)
11. W.N. Plick, M. Krenn, R. Fickler, S. Ramelow, A. Zeilinger, Quantum orbital angular momentum of elliptically symmetric light. Phys. Rev. A **87**, 033806 (2013)
12. H.R. Yang, H.J. Wu, W. Gao, C. Rosales-Guzmán, Z.H. Zhu, Parametric upconversion of Ince-Gaussian modes. Opt. Lett. **45**, 3034–3037 (2020)
13. Z.Y. Bai, D.M. Deng, Q. Guo, Elegant Ince-Gaussian beams in a quadratic-index medium. Chin. Phys. B **20**, 094202 (2011)
14. Y.Q. Xu, G.Q. Zhou, Propagation of Ince-Gaussian beams in uniaxial crystals orthogonal to the optical axis. Eur. Phys. J. D **66**, 59 (2012)
15. E. Robertson, D.G. Pires, K. Dai, J. Free, K. Kimmel, N. Litchinitser, J.K. Miller, E.G. Johnson, Constant-envelope modulation of Ince-Gaussian beams for high bandwidth underwater wireless optical communications. J. Lightwave Techn. **41**, 5209–5216 (2023)
16. M. Bayraktar, Scintillation performance of Ince-Gaussian beam in atmospheric turbulence. Res. Square **31**, 548 (2023)
17. A.E. Siegman, *Lasers* (University Science Books, Sausalito, 1986)
18. E.L. Ince, A linear differential equation with periodic coefficients. Proc. London Math. Soc. **23**, 56–74 (1923)
19. V.M. Dubrovskii, Equations of degree four. Russ. Math. Surv. **28**, 212 (1973)
20. G.A. Baker, P. Graves-Morris, *Padé Approximants* (Cambridge University Press, Cambridge, 1996)
21. E.G. Abramochkin, V.V. Kotlyar, A.A. Kovalev, Structurally invariant higher-order Ince-Gaussian beams and their expansions into the Hermite-Gaussian or Laguerre-Gaussian beams. Appl. Sci. **14**, 1759 (2024)
22. T. Yu, H. Xia, W. Xie, G. Xiao, H. Li, The generation and verification of Bessel-Gaussian beam based on coherent beam combining. Results Phys. **16**, 102872 (2020)
23. C.P. Boyer, E.G. Kalnins, W. Miller, Lie theory and separation of variables: the harmonic oscillator in elliptic coordinates and Ince polynomials. J. Math. Phys. **16**, 512–517 (1975)
24. L.F. Votto, A. Chafiq, G. Gouesbet, L.A. Ambrosio, A. Belafhal, Ince-Gaussian beams in the generalized Lorenz-Mie theory through finite series Laguerre-Gaussian beam shape coefficients. J. Quantit. Spectr. Radiat. Transf. **302**, 108565 (2023)
25. S. Nomoto, A. Goldstein, R. Vyas, S. Singh, Polarization properties of Ince-Gaussian laser beams. J. Opt. Soc. Am. A **34**, 2261–2265 (2017)
26. H. Ahmed, M.A. Ansari, L. Paterson, J. Li, X. Chen, Metasurface for engineering superimposed Ince-Gaussian beams. Adv. Mater. **36**, 2312853 (2024)
27. Y.X. Ren, Z.X. Fang, L. Gong, K. Huang, Y. Chen, R. De Lu, Dynamic generation of Ince-Gaussian modes with a digital micromirror device. J. Appl. Phys. **117**, 133106 (2015)

28. S.-C. Chu, C.-S. Yang, K. Otsuka, Vortex array laser beam generation from a Dove prism-embedded unbalanced Mach-Zehnder interferometer. Opt. Exp. **16**, 19934–19949 (2008)
29. T. Ohtomo, K. Kamikariya, K. Otsuka, S. Chu, Single-frequency Ince-Gaussian mode operations of laser-diode-pumped microchip solid-state lasers. Opt. Exp. **15**, 10705–10717 (2007)
30. U.T. Schwarz, M.A. Bandres, J.C. Gutiérrez-Vega, Observation of Ince-Gaussian modes in stable resonators. Opt. Lett. **29**, 1870–1872 (2004)
31. M.-Y. Wang, J. Tang, H.-J. Wang, Y. Ming, Y. Zhang, G.-X. Cui, Y.-Q. Lu, Generation of second-harmonic Ince-Gaussian beams. Appl. Phys. Lett. **113**, 081105 (2018)
32. E.N. Castañeda, J.C. Guerra-Vázquez, R.R. Alarcón, I. Agha, Q. Zhan, W.N. Plick, Ince-Gauss beams in a turbulent atmosphere: the effect of structural parameters on beam resilience. Opt. Contin. **1**, 1777–1794 (2022)
33. M. Woerdemann, C. Alpmann, C. Denz, Optical assembly of microparticles into highly ordered structures using Ince-Gaussian beams. Appl. Phys. Lett. **98**, 111101 (2011)
34. S. Lopez-Aguayo, J.C. Gutiérrez-Vega, Elliptically modulated self-trapped singular beams in nonlocal nonlinear media: ellipticons. Opt. Express **15**, 18326–18338 (2007)
35. S. Sakpal, G. Milione, M.-J. Li, M. Nouri, H. Shahoei, T. LaFave, S. Ashrafi, D. MacFarlane, Stability of Ince-Gaussian beams in elliptical core few-mode fibers. Opt. Lett. **43**, 2656–2659 (2018)
36. Y. Yu, Y. Chen, C. Wang, J. Wang, Z. Sun, M. Cao, H. Gao, F. Li, Optical storage of Ince-Gaussian modes in warm atomic vapor. Opt. Lett. **46**, 1021–1024 (2021)
37. E. Abramochkin, E. Razueva, V. Volostnikov, General astigmatic transform of Hermite–Laguerre-Gaussian beams. J. Opt. Soc. Am. A **27**, 2506–2513 (2010)
38. J.A. Rodrigo, T. Alieva, M.L. Calvo, Gyrator transform: properties and applications. Opt. Express **15**, 2190–2203 (2007)
39. E. Abramochkin, T. Alieva, J.A. Rodrigo, *Mathematical optics: classical, quantum, and computational methods* (CRC Press, Boca Raton, 2012)
40. E. Abramochkin, V. Volostnikov, Generalized Gaussian beams. J. Opt. A Pure Appl. Opt. **6**, S157–S161 (2004)
41. V.V. Kotlyar, A.A. Kovalev, Orbital angular momentum of paraxial propagation-invariant laser beams. J. Opt. Soc. Am. A **39**, 1061–1065 (2022)
42. V.V. Kotlyar, A.A. Kovalev, A.V. Volyar, Topological charge of a linear combination of optical vortices: topological competition. Opt. Express **28**, 8266–8281 (2020)
43. M.V. Berry, Optical vortices evolving from helicoidal integer and fractional phase steps. J. Opt. A Pure Appl. Opt. **6**, 259–268 (2004)

Chapter 3
New Type of Laguerre-Gaussian Beams

3.1 Product and Squared Laguerre-Gaussian Beams

Among a host of familiar laser beams, most popular and well-understood are Laguerre-Gaussian (LG) beams [1–3]. Initially, these beams were looked upon as intra-cavity modes, but later on out-of-resonator LG beams were generated from Hermite-Gaussian (HG) beams using an astigmatic converter [4]. Particular interest in the LG beams was provoked by Allen's et al. [5] in which the LG beams were found to carry the angular orbital momentum (OAM). Extensively studied beams include generalized LG beams in the form of Hermite–Laguerre-Gaussian beams [6, 7], elegant [8] and elliptic [9] LG beams. These days have seen no signs of waning interest in studying the LG beams thanks to their wide use in telecommunications, micromanipulation, probing atmospheric turbulence, quantum information, and atom cooling. By way of illustration, a comparative analysis of LG beams and Bessel-Gaussian (BG) beams has been conducted [10]. Various approaches to generating LG modes discussed in Refs. [11–13] included the use of a specialized laser utilizing intra-cavity spherical aberration [11], q-plates [12], and a special metasurface [13]. Reciprocal HG-to-LG and LG-to-HG mode conversion was studied in Ref. [14]. Of essential significance is the study of elegant LG beams that have shown outstanding characteristics for many application areas such as optical communications and optical trapping [15]. A method for measuring a topological charge of a partially coherent elegant LG beam has been proposed [16]. The LG beams have formed a basis for developing new types of optical beams with a variety of promising properties. A family of asymmetric LG laser beams has been discussed [17] and a technique for generating high-power asymmetric LG beams has been proposed [18]. Using the LG beams, a vector beam with space dependent transverse polarization has been generated by a method of nonlinear magnetooptic rotation [19]. A new class of composite vortex beams generated by coaxially superimposing LG beams with identical waist location and parameters has been proposed [20]. A new type of a partially coherent beam with a peculiar correlation function, which has been given

V. V. Kotlyar et al., *Helical Laser Beams*, https://doi.org/10.1007/978-3-031-92217-6_3

the name an elliptic correlated Laguerre–Gauss-Shell model, has been theoretically and experimentally studied [21]. In the far field, the intensity pattern of such beams is in the form of an ellipse-shaped ring. The LG beams are of great practical value for optical communication [22–25], micromanipulation [26], and photo-induced atom excitation [27].

In this section, we propose a new type of laser beams generated by multiplying two LG beams, also termed as a product of LG (pLG) beams. We show that the complex amplitude of the pLG beams can be decomposed into a finite sum of conventional LG beams, deriving a relationship for a Fresnel transform of such beams. For some particular cases of the pLG beams, an explicit form of a Fourier transform is deduced.

3.1.1 Theoretical Background

Let us consider a conventional LG beam described in the source plane (z = 0) by a complex amplitude [1]:

$$\mathrm{LG}_{p,\pm m}(r,\varphi)=\exp\left(-\frac{r^2}{2w^2}\pm im\varphi\right)\left(\frac{r}{w}\right)^m L_p^m\left(\frac{r^2}{w^2}\right), \tag{3.1}$$

where (r,φ) are the polar coordinates, w is the waist radius of a Gaussian beam, and $L_m^n(x)$ is a Laguerre polynomial. With Eq. (3.1) describing a modal beam, i.e. the one which preserves its intensity pattern upon free-space propagation, the complex amplitude of a LG beam at any plane z is described by a relationship similar to Eq. (3.1):

$$\begin{aligned}\mathrm{LG}_{p,m}(r,\varphi,z)=&\frac{w}{w(z)}exp\left(\frac{izr^2}{z_0w^2(z)}+i(2p+m+1)\gamma(z)\right)\\&\mathrm{LG}_{p,m}\left(\frac{wr}{w(z)},\varphi\right),\end{aligned} \tag{3.2}$$

where $w(z)=w[1+(z/z_0)^2]^{1/2}$ is the Gaussian beam radius, $\gamma(z)=\arctan(z/z_0)$, $\gamma(z)$ is the Gough phase, $z_0=kw^2$ is the Rayleigh range, and k is the wavenumber of light.

The product of two LG beams in Eq. (3.1) with identical waist radii, also termed as a pLG beam, is given by

$$\begin{aligned}\mathrm{pLG}_{p,q,m,n}(r,\varphi)&=\mathrm{LG}_{p,m}(r,\varphi)\mathrm{LG}_{q,n}(r,\varphi)\\&=\exp\left(-\frac{r^2}{w^2}+i(m+n)\varphi\right)\left(\frac{r}{w}\right)^{m+n}\\&\quad L_p^m\left(\frac{r^2}{w^2}\right)L_q^n\left(\frac{r^2}{w^2}\right).\end{aligned} \tag{3.3}$$

A four-parameter family of beams in Eq. (3.3) presents a generalization of conventional LG beams of Eq. (3.1), as it becomes identical to single-ring LG beams at $p = q = 0$:

$$\begin{aligned} \mathrm{pLG}_{0,0,m,n}(r,\varphi) &= \exp\left(-\frac{r^2}{w^2} + i(m+n)\varphi\right)\left(\frac{r}{w}\right)^{m+n} \\ &= 2^{-(n+m)/2}\mathrm{LG}_{0,m+n}\left(\sqrt{2}r,\varphi\right). \end{aligned} \tag{3.4}$$

From (3.4), the topological charge (TC) of a pLG beam can be found as a sum of TCs of two constituent LG beams. The power-normalized orbital angular momentum (OAM) of the pLG beam is also found as a sum of two constituent LG beams, i.e. equals $m + n$. Then, we proceed to derive a relationship for the Fresnel transform of function (3.4). The pLG beam can be decomposed into a finite sum of conventional LG beams:

$$L_p^m(x)L_q^n(x) = \sum_{k=0}^{p+q} C_k L_k^{m+n}(2x), \tag{3.5}$$

where $C_k = 2^{q-k}P_{p+q-k}^{(k+m-q,k+n-p)}(0)P_q^{(k-q,p-k)}(0)$ and $P_\alpha^{\mu,\nu}(0)$ denote Jacobi polynomials at zero [28]. From (3.5), the amplitude of the pLG beam in (3.4) is given by a finite sum of conventional LG beams:

$$\begin{aligned} \mathrm{pLG}_{p,q,m,n}(r,\varphi) &= \exp\left(-\frac{r^2}{w^2} + i(m+n)\varphi\right)\left(\frac{r}{w}\right)^{m+n} \\ &\quad \sum_{k=0}^{p+q} C_k L_k^{m+n}\left(\frac{2r^2}{w^2}\right) \\ &= 2^{-(m+n)/2}\sum_{k=0}^{p+q} C_k \mathrm{LG}_{k,m+n}(\sqrt{2}r,\varphi). \end{aligned} \tag{3.6}$$

The relationship in (3.6) enables the amplitude of field (3.4) to be derived at any distance z:

$$\begin{aligned} \mathrm{pLG}_{p,q,m,n}(r,\varphi,z) = &\frac{w}{2^{(n+m)/2}w(z)}\exp\left(-\frac{r^2}{2w^2(z)} + \frac{ikr^2}{2R(z)} + i(m+n)\varphi\right) \\ &\times\left(\frac{r}{w(z)}\right)^{n+m}\sum_{s=0}^{p+q} C_s L_s^{n+m}\left(\frac{r^2}{w^2(z)}\right) \\ &\exp(i(2s+m+n+1)\gamma(z)). \end{aligned} \tag{3.7}$$

With the constituent beams in superposition (3.7) having different Gough phase, we can infer that the pLG beam of Eq. (3.4) does not preserve its structure upon free-space propagation. However, with the transverse intensity pattern of beam (3.7) given by a set of concentric intensity rings whose maximum number is $p + q + 1$, changes in the intensity pattern upon free-space propagation of beam (3.7) are limited to the inter-ring energy flow.

It would be of interest to analyze a particular case of pLG beams whose constituent Laguerre polynomials are identical. In the source plane, the beam, which has been given the name 'squared' LG beam, or (LG)^2 beam, is given by, instead of (3.4)

$$\begin{aligned}\mathrm{pLG}_{p,p,m,m}(r,\varphi) &= \left[\mathrm{LG}_{p,m}(r,\varphi)\right]^2 \\ &= \exp\left(-\frac{r^2}{w^2} + i2m\varphi\right)\left(\frac{r}{w}\right)^{2m}\left[L_p^m\left(\frac{r^2}{w^2}\right)\right]^2.\end{aligned} \tag{3.8}$$

Similar to the general pLG beam in (3.4), beam (3.8) is not modal and does not preserve the intensity structure upon free-space propagation. At the same time, unlike beam (3.4), in the far field beam (3.8) preserves its structure, being Fourier-invariant:

$$\begin{aligned}F_f\left\{\mathrm{pLG}_{pp,mm}(r,\varphi)\right\} &= \left(\frac{iz_0}{f}\right)(-1)^{m+1}\mathrm{pLG}_{pp,mm}(\rho,\theta) \\ &= \left(\frac{iz_0}{f}\right)(-1)^{m+1}\exp\left(2im\theta - y^2\right)(y)^{2m}\left[L_p^m\left(y^2\right)\right]^2,\end{aligned} \tag{3.9}$$

where $y = kw\rho/f$, f is the focal length of a spherical lens, and (ρ, θ) are the polar coordinates in the Fourier plane. A comparison of relationships for complex amplitudes in the source plane, Eq. (3.8), and at the focus of a spherical lens, Eq. (3.9), shows that they are identical up to a constant. In the Fresnel diffraction zone, the (LG)^2 beam is described by a finite superposition of conventional LG beams, which are similar to Eq. (3.7) but have different coefficients:Γ

$$\begin{aligned}\mathrm{pLG}_{pp,mm}(r,\varphi,z) = {} & \left(\frac{iz_0}{z}\right)(-1)^{m+1}\frac{\Gamma(m+p+1)}{\pi m!\sigma^{2m+1}}(y)^{2m} \\ & \exp\left(\frac{ikr^2}{2z} + 2im\varphi\right)\exp\left(-\frac{y^2}{\sigma}\right) \\ & \times\sum_{s=0}^{p}\frac{(-1)^s}{(p-s)!}\frac{\Gamma(p-s+1)\Gamma(s+1/2)}{\Gamma(m+s+1)} \\ & \left(\frac{\sigma-2}{2}\right)^{2s}L_{2s}^{2m}\left(\frac{y^2}{\sigma(2-\sigma)}\right),\end{aligned} \tag{3.10}$$

where $y = kw\rho/z$, $\sigma = 1 - iz_0/z$, $\Gamma(x)$ is the gamma-function. In the general case, Eq. (3.10) is consistent with Eq. (3.7) because in both cases the complex amplitude is expressed via a finite sum of conventional LG beams. The difference between Eqs. (3.7) and (3.10) is that the latter only contains LG beams with even radial indices. Equation (3.10) also suggests that at $p = 0$ (zero-valued radial index) the intensity pattern of the LG beam is a single ring, as $L_0^n(x) = 1$, so in the sum in Eq. (3.10) only the first term is retained, meaning that beam (3.1) with a squared amplitude preserves upon free-space propagation.

Another particular case of pLG beams in Eq. (3.4) is obtained by considering the product of two LG beams with even TC and tailored values of azimuthal numbers in the Laguerre polynomials:

$$\begin{aligned} \mathrm{pLG}_{p,q,n-m,n+m}(r,\varphi) &= \exp\left(-\frac{r^2}{w^2} + i2n\varphi\right) \\ &\times \left(\frac{r}{w}\right)^{2n} L_p^{n-m}\left(\frac{r^2}{w^2}\right) \\ &L_q^{n+m}\left(\frac{r^2}{w^2}\right), \quad n \geq m \geq 0. \end{aligned} \tag{3.11}$$

The complex amplitude of beam (3.11) at the focus of a spherical lens (in the Fourier plane) is given by an explicit relationship:

$$\begin{aligned} F_f\left\{\mathrm{pLG}_{pq,(n-m)(n+m)}(r,\varphi)\right\} &= \left(\frac{iz_0}{f}\right)(-1)^{m+1}\exp\left(2in\theta - y^2\right) \\ &\times (y)^{2n} L_p^{n-m+p-q}\left(y^2\right) L_q^{n+m-p+q}\left(y^2\right). \end{aligned} \tag{3.12}$$

From Eq. (3.12), the product of two beams in Eqs. (3.4) and (3.11) is seen to be Fourier-invariant at $p = q$. From (3.11) and (3.12), at $p = q$ and $m = 0$, conversion of a pLG beam into an $(\mathrm{LG})^2$ beam is seen to occur.

We can derive a relationship for the 'energy' of the pLG beam, using it when normalizing such beams:

$$\begin{aligned} W &= \int_0^\infty \int_0^{2\pi} \left|\mathrm{pLG}_{p,q,m,n}(r,\varphi)\right|^2 r dr d\varphi \\ &= \frac{\pi w^2}{4}\{(m+p)!(n+q)!\}^2 \\ &\sum_{k=0}^{\min(2m+p,2n+q)} \frac{\left\{P_{m+p}^{m-k,k-m-p}(0) P_{n+q}^{n-k,k-n-q}(0)\right\}^2}{(2m+p-k)!(2n+q-k)!k!^2}. \end{aligned} \tag{3.13}$$

3.1.2 Numerical Simulation

We have carried out numerical simulation of focusing the $(\mathrm{LG})^2$ beam by a spherical lens using our own MATLAB scripts. The initial field was represented as:

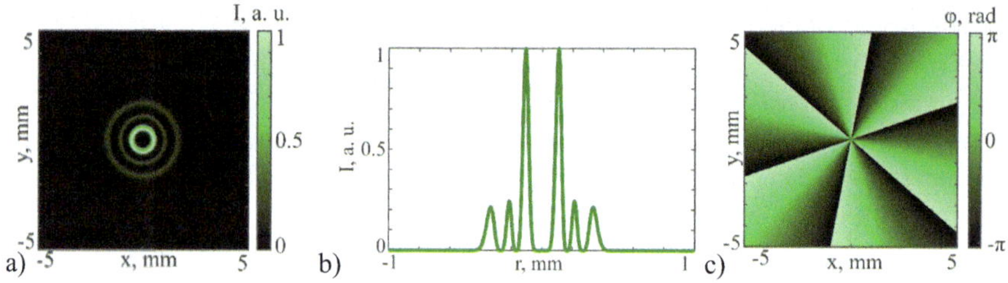

Fig. 3.1 Initial $(LG)^2$ beam: 2D intensity distribution (**a**); intensity cross section along the radius (**b**); 2D phase distribution (**c**)

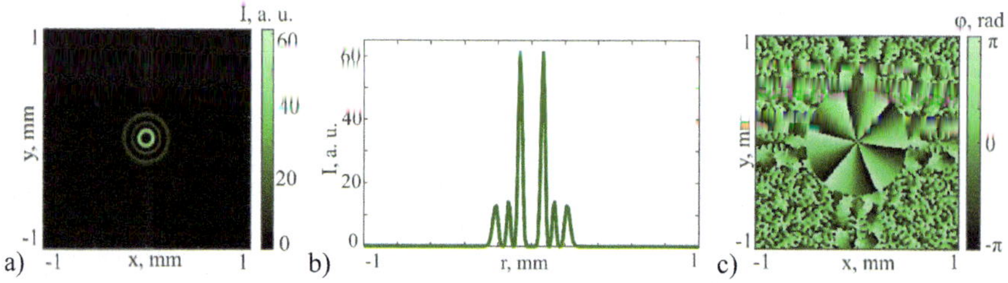

Fig. 3.2 Field at the focus of a spherical lens after focusing the initial beam from Fig. 3.1: 2D intensity distribution (**a**); intensity cross section along the radius (**b**); 2D phase distribution (**c**)

$$(LG)^2_{n,m}(r,\varphi) = \exp\left(-\frac{r^2}{w^2} + i2n\varphi\right)\left(\frac{r}{w}\right)^{2|n|}\left[L_m^{|n|}\left(\frac{r^2}{w^2}\right)\right]^2. \tag{3.14}$$

Figure 3.1 shows the initial intensity and phase distribution for $(LG)^2$ with the following parameters: $\lambda = 532$ nm, $w = 0.5$ mm, $n = 3$, $m = 2$.

Focusing with a spherical lens is described by the Fourier transform. The simulation results for this beam in the focus are shown in Fig. 3.2.

Figures 3.1 and 3.2 differ only by a constant and clearly demonstrate the Fourier invariance of $(LG)^2$ beams proved in the first section. After that we consider the Fresnel diffraction, which is represented by the following integral:

$$E(x,y,z) = -\frac{i}{\lambda}\frac{e^{ikz}}{z}\int\int_{-\infty}^{+\infty} E(x\prime,y\prime,0)$$

$$\exp\left\{\frac{ik}{2z}\left[(x-x\prime)^2 + (y-y\prime)^2\right]\right\}dx\prime dy\prime. \tag{3.15}$$

For its numerical calculation, the convolution theorem and the properties of the Fourier transform are used, which make it possible to calculate the integral (3.15) by using direct and inverse Fourier transforms. The calculation program was also implemented in the MATLAB package. The simulation results at different distances are shown in Figs. 3.3, 3.4, and 3.5.

Figure 3.3 shows the intensity (a), its cross section (b) and the phase (c) of the $(LG)^2$ beam shown in Fig. 3.1, but at a half of the Rayleigh length. It can be seen in Fig. 3.3 that instead of three intensity rings (Fig. 3.1), a fourth intensity ring appears in the beam. In addition, the brightest ring is no longer the first one (Fig. 3.1). It is the second, instead.

Figure 3.4 shows the intensity (a), its cross section (b), and phase (c) of the same beam (Fig. 3.1), but at the Rayleigh distance. It can be seen in Fig. 3.4 that the

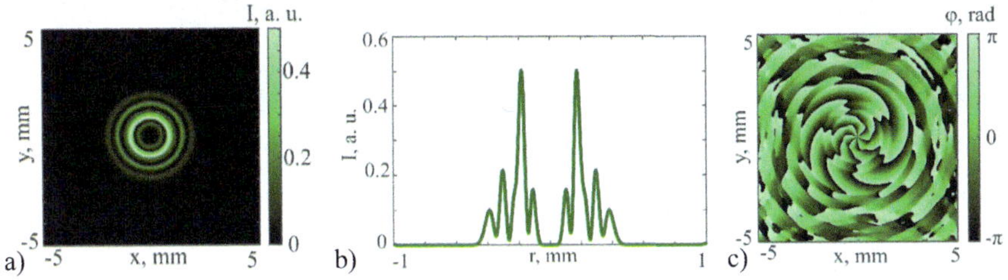

Fig. 3.3 The Fresnel-transformed field of the initial $(LG)^2$ beam (3.14) at a distance $z = z_0/2$ ($z_0 \approx 1.476$ m): 2D intensity distribution (**a**); intensity cross section along the radius (**b**); 2D phase distribution (**c**)

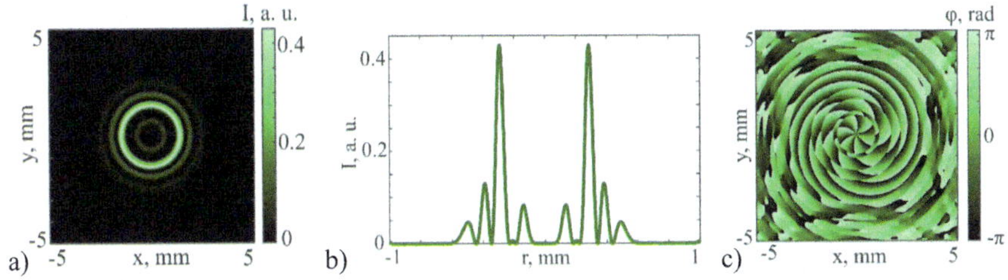

Fig. 3.4 The transformed field of the initial $(LG)^2$ beam (3.14) at a distance $z = z_0$ ($z_0 \approx 1.476$ m): 2D intensity distribution (**a**); intensity cross section along the radius (**b**); 2D phase distribution (**c**)

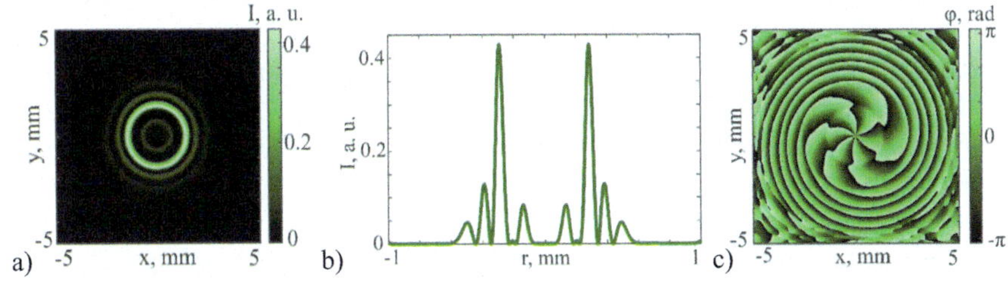

Fig. 3.5 The transformed field of the initial $(LG)^2$ beam (3.14) at a distance $z = 2z_0$ ($z_0 \approx 1.476$ m): 2D intensity distribution (**a**); intensity cross section along the radius (**b**); 2D phase distribution (**c**)

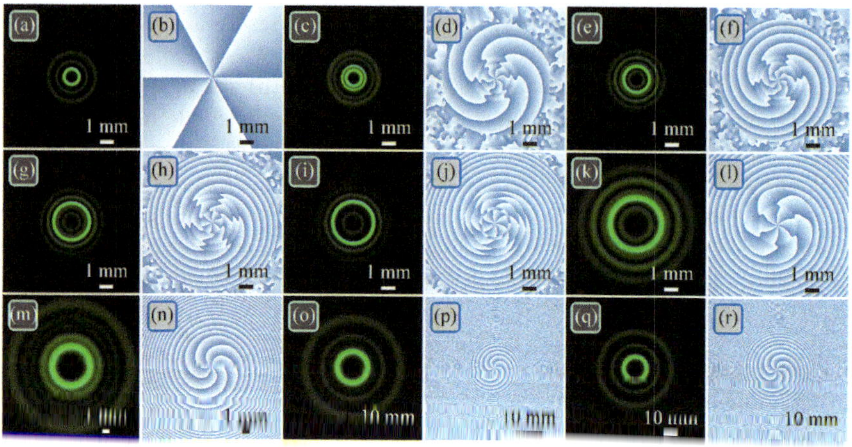

Fig. 3.6 The transformed field of the initial $(LG)^2$ beam (3.14) at distances $z = 0$ (**a**, **b**), $z = z_0/4$ (**c**, **d**), $z = z_0/2$ (**e**, **f**), $z = 3z_0/4$ (**g**, **h**), $z = z_0$ (**i**, **j**), $z = 2z_0$ (**k**, **l**), $z = 5z_0$ (**m**, **n**), $z = 10z_0$ (**o**, **p**), $z = 15z_0$ (**q**, **r**): intensity (**a**, **c**, **e**, **g**, **i**, **k**, **m**, **o**, **q**) and phase (**b**, **d**, **f**, **h**, **j**, **l**, **n**, **p**, **r**)

beam has four bright rings, but the energy distribution between them differs from that presented in Fig. 3.3.

Figure 3.5 shows the same as Figs. 3.3 and 3.4, but at a distance of two Rayleigh lengths. There are still 4 rings in the intensity distribution.

Figure 3.6 shows together the intensity distributions of the same beam as in Figs. 3.1, 3.2, 3.3, 3.4, and 3.5. It is seen that the cross section of the beam consists of three bright rings again at a distance of 10 Rayleigh lengths (i.e. in the far field). It should be noted that the brightest ring is the first one from the center. And at a distance of 15 Rayleigh lengths, the intensity distribution coincides with the initial intensity distribution, differing only in scale. Thus, numerical simulation confirms all theoretical predictions.

Figure 3.7 depicts numerically simulated intensity and phase patterns from the standard LG beams (3.1) of two different orders and from the pLG beam (3) of the same two orders, shown in the source plane and at the Rayleigh range. We note that with the waist radius of beam (3.1) taken to be $2^{1/2}w$, and not w, the Rayleigh range of beam (3.1) is twice that of beam (3.3). Shown in Fig. 3.7 are intensity and phase patterns at the Rayleigh range for beam (3.3), i.e. at $kw^2/2$. The patterns have been numerically simulated using Eqs. (3.1) and (3.3) in the source plane and using a Fresnel transform at the Rayleigh range.

From Fig. 3.7, the intensity profiles are seen to be preserved upon free-space propagation for the both conventional LG beams, while varying for the pLG beam. In particular, in the source plane, the first (from the center) intensity ring is brightest, with the second ring becoming brightest at the Rayleigh range. According to Eq. (3.3), the beam is supposed to have $p + q + 1 = 6$ bright intensity rings. However, just four and five intensity rings can, respectively, be seen in Fig. 3.7i and k. Meanwhile, the intensity profile in Fig. 3.7i and phase pattern in Fig. 3.7k confirm that there are

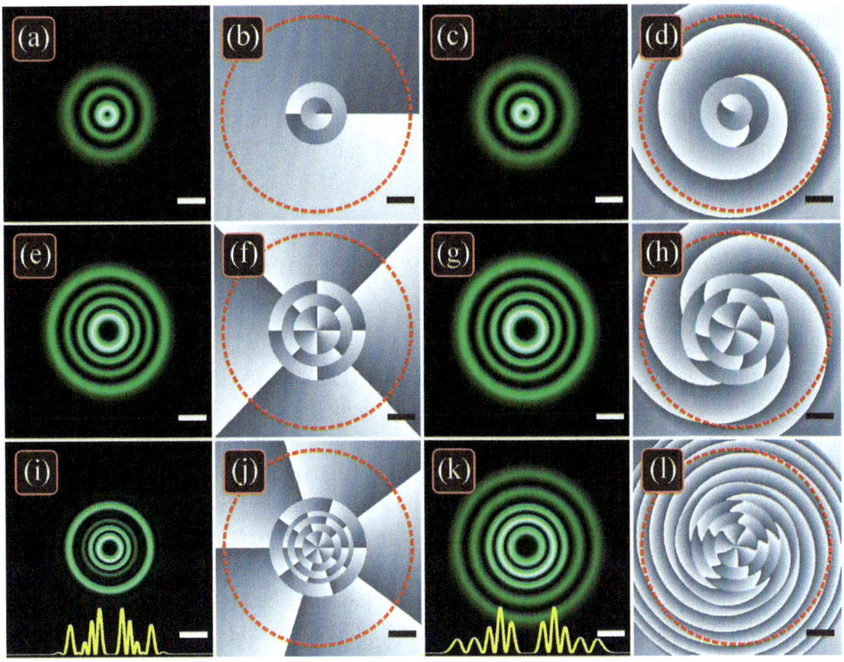

Fig. 3.7 Patterns of intensity (columns 1 and 3) and phase (columns 2 and 4, dark—0, white—2π) from conventional LG beams (3.1) of the orders $(p, m) = (2, 1)$ (row 1) and $(q, n) = (3, 4)$ (row 2) and from a pLG beam (3.3) of the order $(p, q, m, n) = (2, 3, 1, 4)$ (row 3) in the source plane $z = 0$ (columns 1 and 2) at the Rayleigh range $z = z_0$ (columns 3 and 4) for the following parameters: wavelength $\lambda = 532$ nm, waist radius $w = 0.5$ mm. The scale bar in all pictures is 1 mm. The topological charge was measured along a dashed circle in the phase patterns. Shown in the bottom row in this figure (**i, k**) are intensity profiles

six bright rings and five intermitting dark rings (where a phase jump by π occurs). In Fig. 3.7k, the rings are fused, reducing the total number to five.

As predicted by Eq. (3.3), TC of beam (3.3) is found as a sum of TCs of Eq. (3.1): $1 + 4 = 5$. This can be seen from the phase pattern: in Fig. 3.7c, d, TC = 1 (a single phase jump by 2π along the dashed circle), in Fig. 3.7f, h TC = 4 (four phase jumps by 2π along the circle), and in Fig. 3.7j, l TC = 5 (five phase jumps along the circle).

With the Rayleigh range of the Laguerre beams (3.1) being twice as large, following the propagation over a distance of $z = kw^2/2$, these diverge not by a factor of $2^{1/2} \approx 1.41$, but by a factor of just $(5/4)^{1/2} \approx 1.12$ (Fig. 3.7a, c, e, g). In the meantime, beam (3.3) expands by a factor of just about 1.5.

Summing up, we have introduced a four-parameter family of vortex beams overlapping well-known LG beams. The new beams represent a product of two different LG beams with the same radius of waist and can also be termed as a product of LG (pLG) beams [29, 30]. If in a pLG beam both Laguerre polynomials have the same indices, such a beam may be referred to as a 'squared' LG beam, or $(LG)^2$ beam. The proposed pLG beams have been expressed as superposition of a finite sum of conventional LG beams. For the $(LG)^2$ beams, an explicit Fourier transform has been derived. A particular case of pLG beams whose Laguerre polynomials are described by specially tailored azimuthal indices n-m and $n + m$ has been analyzed and their Fourier transform has been deduced in an explicit form. The pLG beams are promising for optical communication applications [22, 24].

3.2 Laguerre-Gaussian Beams with an Increased Dark Area and Autofocusing

During many years, vortex Laguerre-Gaussian (LG) beams attract a stable interest due to their wide practical applications in optical trapping [31], wireless telecommunications [23], atmosphere probing [32], quantum informatics [19], atom cooling [33], optical microscopy of single molecules [34] and quantum dots [35]. Besides, the LG beams have been very thoroughly studied theoretically. For instance, possibility of the orbital angular momentum (OAM) of light was at first shown right for the LG beams [5]. There are known many modifications of the LG beams, like, for example, elegant LG beams [36], Hermite–Laguerre-Gaussian beams [7], asymmetric LG beams [17], products of LG beams [30]. The LG beams can be generated by lasers [11, 37–39], mode converters [4], and using a spatial light modulator [40]. However, despite the long history of the LG beams, their potential is still not exhausted.

In this section, we investigate a novel LG beam with its topological charge being equal to the difference between the azimuthal (upper) and radial (lower) indices of the associated Laguerre polynomial, whereas the radial polar coordinate is raised to a power equal to the sum of these indices. Thus, in contrast to the conventional LG beams, the topological charge of the studied beam is not equal to the power of the radial coordinate. On the one hand, it violates the modal propagation (shape invariance) of such a beam in free space. On the other hand, the beam discussed demonstrates interesting properties: autofocusing and ability to control the diameter of the central dark intensity spot without changing the topological charge of the beam.

3.2.1 *Fourier-Invariant Laguerre-Gaussian Beams with an Increased Dark Area*

Here, we consider a coherent paraxial monochromatic light field with the following complex amplitude in the initial plane ($z = 0$):

$$E_{n,m}(r,\varphi) = \exp\left(-\frac{r^2}{w^2} + i(m-n)\varphi\right)\left(\frac{r}{w}\right)^{m+n} L_n^m\left(\frac{r^2}{w^2}\right), \tag{3.16}$$

where (r, φ) are the polar coordinates, w is the waist radius of the Gaussian beam, $L_n^m(x)$ is the associated Laguerre polynomial. Using standard LG modes

$$\mathrm{LG}_{p,\pm\ell}(\mathbf{r}) = \exp\left(-\frac{r^2}{w^2} \pm i\ell\varphi\right)\left(\frac{r}{w}\right)^{\ell} L_p^{\ell}\left(\frac{2r^2}{w^2}\right), \tag{3.17}$$

the beam (3.16) can be written as a product of two of them:

$$E_{n,m}(\mathbf{r}) = 2^{(n+m)/2}\mathrm{LG}_{n,m}\left(\frac{\mathbf{r}}{\sqrt{2}}\right)\mathrm{LG}_{0,-n}\left(\frac{\mathbf{r}}{\sqrt{2}}\right). \tag{3.18}$$

The structure of its light rings does not conserve under the beam propagation in free space. However, as we will see later, the beam from Eq. (3.16) has some interesting properties useful in practice: (i) the beam from Eq. (3.16) is Fourier-invariant, (ii) it is also autofocusing at the Rayleigh distance from the waist, and (iii) it has an increased dark area near the optical axis, i.e. the diameter of the central dark intensity spot can be tuned (increased) without changing the topological charge.

Equation (3.16) indicates that the topological charge of the optical vortex equals $m - n$, and the orbital angular momentum of the beam (3.16), normalized to the beam power, is also equal to $m - n$. Initially, the beam has $n + 1$ light rings. However, since the beam (3.16) does not conserve its structure on propagation, the number of rings can change. Thus, $n + 1$ is the maximal number of rings of the beam (3.16). Differentiating the intensity of the field (3.16), $I_{n,m}(r) = \left|E_{n,m}(\mathbf{r})\right|^2$, with respect to r and equating the derivative to zero, we find the intensity maxima [minima are evidently zeroes of the beam (3.1)]:

$$\left(s - \tfrac{1}{2}[m+n]\right)L_n^m(s) + sL_{n-1}^{m+1}(s) = 0, \tag{3.19}$$

where $s = (r/w)^2$. Equation (3.19) is an algebraic equation of the $(n + 1)$th order and can be analytically solved only for small values of n only. For instance, since at $n = 1$ the Laguerre polynomial is equal to $L_1^m(y) = -y + (m + 1)$, Eq. (3.17) allows obtaining the radius of the first light ring: $r \sim w\sqrt{m/2}$. In the general case, the radius of the first light ring depends on the radial index n, as follows from Eq. (3.19). Below, we consider a particular case of Eq. (3.16), when the topological charge of the beam is nonnegative. Replacing $m = n + \ell$, , the beam (3.16) can be rewritten as follows,

$$E_{n,n+\ell}(\mathbf{r}) = \exp\left(-\frac{r^2}{w^2} + i\ell\varphi\right)\left(\frac{r}{w}\right)^{2n+\ell} L_n^{n+\ell}\left(\frac{r^2}{w^2}\right), \tag{3.20}$$

where $\ell \geq 0$. Equation (3.20) shows that, without changing the topological charge (ℓ = const), the effective diameter of the dark area near the optical axis can be changed by varying only the index n. When n increases, the dark low-intensity domain near the optical axis becomes larger, whereas decreasing n reduces this domain.

The beam (3.16) does not keep the intensity shape during propagation in the Fresnel diffraction zone. However, its has an invariant Fourier image. Namely,

$$\frac{-iz_0}{\pi z}\iint\limits_{\mathbb{R}^2}\exp\left(\frac{iz_0}{z}\cdot\left|\frac{\mathbf{r}-\boldsymbol{\rho}}{w}\right|^2-\frac{iz_0}{f}\cdot\left|\frac{\boldsymbol{\rho}}{w}\right|^2\right)E_{n,m}(\boldsymbol{\rho})\frac{d^2\boldsymbol{\rho}}{w^2}\Bigg|_{z=f}$$

$$=\frac{-iz_0}{\pi f}\exp\left(\frac{iz_0}{f}\cdot\left|\frac{\mathbf{r}}{w}\right|^2\right)\iint\limits_{\mathbb{R}^2}\exp\left(-\frac{2iz_0}{f}\cdot\frac{\langle\mathbf{r},\boldsymbol{\rho}\rangle}{w^2}\right)E_{n,m}(\boldsymbol{\rho})\frac{d^2\boldsymbol{\rho}}{w^2}$$

$$=(-i)^{m-n+1}\cdot\frac{z_0}{f}\exp\left(\frac{iz_0}{f}\cdot\left|\frac{\mathbf{r}}{w}\right|^2\right)E_{n,m}\left(\frac{z_0}{f}\mathbf{r}\right). \tag{3.21}$$

where $z_0 = kw^2/2$ is the Rayleigh distance, $\boldsymbol{\rho} = (\xi, \eta) = (\rho\cos\theta, \rho\sin\theta)$ is a 2D vector, f is the focal length of the spherical lens that performs the Fourier transform. It is seen from Eq. (3.21) that the complex amplitude of the light field in the lens focus is equal to the amplitude of the light field (3.16) up to scale and a constant multiplier.

3.2.2 *Fresnel Transform of the Laguerre-Gaussian Beam with an Increased Dark Area*

To evaluate the Fresnel transform of the initial light field (3.3), we apply the expansion:

$$r^{2n}L_p^{n+l}(r^2)L_q^n(r^2)=\sum_{k=0}^{p+q+n}C_kL_k^l(2r^2), \tag{3.22}$$

which may be rewritten in terms of LG modes:

$$\exp\left(-\frac{r^2}{w^2}+i\ell\varphi\right)\left(\frac{r}{w}\right)^{2n+\ell}L_p^{n+\ell}\left(\frac{r^2}{w^2}\right)L_q^n\left(\frac{r^2}{w^2}\right)=\sum_{k=0}^{p+q+n}C_k\mathrm{LG}_{k,\ell}(\mathbf{r}). \tag{3.23}$$

The coefficients $C_k = C_k(p, q, n, \ell)$ are as follows:

$$C_k=\frac{(p+n+\ell)!k!}{2^{p+q+n+\ell+1}p!(k+\ell)!}\cdot[\![u^k]\!]\{(1-u)^{q+n}(1+u)^p\}$$
$$\times[\![v^q]\!]\{(1-v)^{k+\ell}(1+v)^{p+q+n-k}\}. \tag{3.24}$$

Here, we use the notation proposed in [19]. Namely, if $A(t)$ is any power series $\sum_k a_kt^k$, then $[\![t^k]\!]A(t)$ denotes the coefficient of t^k in $A(t)$. As result, both $[\cdot]$

expressions in Eq. (3.24) are integer numbers and may be written in terms of Jacobi polynomials:

$$[\![t^n]\!]\{(1-t)^a(1+t)^b\} = (-2)^n P_n^{(a-n,b-n)}(0) = \sum_{j=\max(0,n-b)}^{\min(n,a)} (-1)^j \binom{a}{j}\binom{b}{n-j}. \tag{3.25}$$

Substituting $p = n$ and $q = 0$ into Eq. (3.24) we obtain the expansion of the beam (3.18) on the base of LG modes:

$$E_{n,n+\ell}(\mathbf{r}) = \sum_{k=0}^{2n} C_k \mathrm{LG}_{k,\ell}(\mathbf{r}) \tag{3.26}$$

with

$$C_k = C_k(n, 0, n, \ell) = \frac{(2n+\ell)!k!}{2^{2n+\ell+1}n!(k+\ell)!} \cdot [\![u^k]\!](1-u^2)^n. \tag{3.27}$$

Since $[\![u^k]\!](1-u^2)^n = 0$ for odd values of k, the expansion (3.27) has only even terms:

$$E_{n,n+\ell}(\mathbf{r}) = \sum_{k=0}^{n} C_{2k} \mathrm{LG}_{2k,\ell}(\mathbf{r}), \quad C_{2k} = \frac{(2n+\ell)!(2k)!}{2^{2n+\ell+1}n!(2k+\ell)!} \cdot (-1)^k \binom{n}{k}. \tag{3.28}$$

According to Eq. (3.28), the light field (3.16) is a superposition of a finite number of the conventional LG beams. Therefore, the complex amplitude of the light field (3.18) can be obtained at an arbitrary propagation distance z from the initial plane. Since

$$\mathrm{LG}_{p,\pm\ell}(\mathbf{r}, z) = \frac{w}{w(z)} exp\left(\frac{izr^2}{z_0 w^2(z)} + i(2p+\ell+1)\psi(z)\right)\mathrm{LG}_{p,\pm\ell}\left(\frac{w\mathbf{r}}{w(z)}\right), \tag{3.29}$$

then

$$\begin{aligned} E_{n,n+\ell}(\mathbf{r}, z) &= \sum_{k=0}^{n} C_{2k} \mathrm{LG}_{2k,\ell}(\mathbf{r}, z) \\ &= \frac{w}{w(z)} exp\left(\frac{izr^2}{z_0 w^2(z)}\right) \\ &\quad \sum_{k=0}^{n} e^{i(4k+\ell+1)\psi(z)} C_{2k} \mathrm{LG}_{2k,\ell}\left(\frac{w\mathbf{r}}{w(z)}\right), \end{aligned} \tag{3.30}$$

where $w(z) = w\left(1 + z^2 / z_0^2\right)^{1/2}$ is the Gaussian beam radius at the distance z, $z_0 = kw^2 / 2$ is the Rayleigh distance, k is the wavenumber of light, $\psi(z) = \arctan(z / z_0)$ is the Gouy phase.

If we recast Eq. (3.30) and write all s-independent multipliers out of the sum, then the remained sum is given by

$$\sum_{s=0}^{2n} C_s \exp(2is\psi(z)) L_s^l\left(\frac{2r^2}{w^2(z)}\right). \tag{3.31}$$

At $z = 0$, the Gouy phase in Eq. (3.31) is zero and the complex amplitude (3.30) coincides with the initial amplitude (3.20). In the far field, when $z >> z_0$, the Gouy phase equals $\pi/2$ and the exponents in Eq. (3.31) become equal to $(-1)^s = 1$, since $s = 2r$. Thus, the sums (3.31) are coinciding at $z = 0$ and at $z >> z_0$. This again proves that the light field from Eq. (3.16) in the far field coincides with the field in the waist. At $z = z_0$, the Gouy phase $\psi(z)$ equals $\pi/4$ and the exponents in Eq. (3.31) are equal to $\exp(i2s\psi(z_0)) = i^s$, or, since only the terms with even s are nonzero in Eq. (3.31), $i^s = (-1)^r, \quad r = 0, 1, 2, ..., n$. Thus, all the terms in the sum are real-valued and of alternating signs.

As seen from Eq. (3.16), the radial (lower) index of the Laguerre polynomial is equal to n, and, therefore, the Laguerre polynomial has also n roots and the LG beam from Eq. (3.16) has $n + 1$ light rings. However, the maximal radial index of the Laguerre polynomial in Eq. (3.31) is equal to $2n$. Thus, it may seem that the sum (3.31) has also $2n$ zeros (roots). But it is not so. Since only the even terms are nonzero in the sum, after replacing the argument x of the Laguerre polynomial ($x^2 = y$), the sum (3.31) with respect to the argument y has the maximal power n and thus has the maximal number of roots equal to n.

Thus, in this Section, applying the Fresnel transform, we obtained the field distribution of the LG beam (3.16) in an arbitrary transverse plane. To do this, we decomposed the light field into a superposition of the conventional LG beams. Strictly speaking, the conventional LG beams constitute a full basis and an arbitrary light field can in principle be expanded into a series of the conventional LG beams. However, our expansion contains only a finite number of terms, the expansion coefficients are explicitly derived [Eq. (3.28)], and all the constituent LG beams have the same azimuthal index (i.e. a vortex factor of the same order), so that the beam (3.16) contains the central optical vortex which on propagation is not split and does not move from the center. Such a decomposition remains the topological charge of the beam (3.16) predictable, but makes its radial distribution more flexible.

3.2.3 Numerical Simulation

In this section, using the Fresnel transform of the LG beam from Eq. (3.16), we compute the transverse intensity and phase distributions at different distances from the initial (waist) plane. We use the following parameters: wavelength $\lambda = 532$ nm, waist radius of the Gaussian beam $w = 0.5$ mm, azimuthal (upper) and radial (lower) indices of the associated Laguerre polynomial are respectively $m = 4$ and $n = 3$. Figure 3.8 illustrates the intensity (columns 1 and 3) and phase (columns 2 and 4) distributions of the beam from Eq. (3.16) at the following distances from the waist z/z_0: 0 (Fig. 3.8a, b), 1/4 (Fig. 3.8c, d), 1/2 (Fig. 3.8e, f), 3/4 (Fig. 3.8g, h), 1 (Fig. 3.1i, j), 1.5 (Fig. 3.8k, l), 2 (Fig. 3.8m, n), 3 (Fig. 3.8o, p), 4 (Fig. 3.8q, r), 10 (Fig. 3.8s, t).

According to Fig. 3.8, at the Rayleigh distance z_0 from the waist, a single light ring is generated with its radius nearly equal to 0.4 mm. This is equivalent to focusing a Gaussian beam with the waist radius of 0.5 mm by a spherical lens with очень слабой the numerical aperture nearly NA = $1.41w/z_0 = 0.0005$. Увеличивая радиальный индекс n, можно в несколько раз увеличить NA. Заметим, что небольшая числовая апертура фокусировки приводит к большой глубине резкости, так как deep of focus (DOF) обратно пропорционален квадрату числовой апертуры. We note также that in the initial plane the first light ring has a radius almost twice as large as the waist radius and equal to nearly 0.9 mm. The conventional LG beam with the waist radius of 0.5 mm would have such a radius at the topological charge of 3, whereas the LG beam from Eq. (3.1) has the topological charge $m - n = 4 - 3 = 1$. Therefore, modeling confirms that the beams (3.1) have an enlarged dark area in the initial plane (Fig. 3.1a) and at the Rayleigh distance a focal spot is formed in the form of a narrow light ring (Fig. 3.1i) almost without side lobes. Although, to be precise, focusing (maximum intensity on the ring) takes place at a distance of 0.75 z_0 (Fig. 3.1g) and 2.38 times greater than the maximum intensity on the ring in the initial plane (Fig. 3.1a). At the Rayleigh distance, the intensity is slightly less (Fig. 3.1i) and 2.37 times greater than in the initial plane. It can also be seen from Fig. 3.1 that the far field intensity (Fig. 3.1s) coincides with the initial intensity (Fig. 3.1a). This confirms the Fourier invariance property (3.4) of beams (3.1). As to the phase evolution, it is seen in Fig. 3.1 that in all transverse planes the phase contains a $(m - n)$th-order singularity in the center, since all the terms in Eq. (3.28) are proportional to $\exp(il\varphi)$. In addition, the vortex phase rotates on propagation and the wavefront acquires an additional parabolic shape due to converging and diverging.

Figure 3.1 was obtained by the numerical Fresnel transform, implemented as a convolution with using the fast Fourier transforms (implemented in the SciPy library for the Python language). To confirm the expansion (3.23) into a superposition of the conventional LG beams, Fig. 3.2 illustrates the same patterns as in Fig. 3.1e, f, o, p, but computed analytically, via the Laguerre polynomials and the expansion coefficients (3.23). It is seen that Fig. 3.2a–d are quite identical to Fig. 3.1e, f, o, p, respectively (Fig. 3.9).

Now when Eq. (3.28) was tested, we can use Eq. (3.30) for computing the longitudinal intensity distributions of the beam from Eq. (3.16) and investigate how the beam transforms when the radial and azimuthal indices vary. Figure 3.10 illustrates these distributions for the following parameters: wavelength $\lambda = 532$ nm, waist radius $w = 0.5$ mm, radial and azimuthal indices (n, m) are respectively (0, 1) (Fig. 3.10a), (0, 4) (Fig. 3.10b), (3, 4) (Fig. 3.10c), (3, 7) (Fig. 3.10d), (6, 7) (Fig. 3.10e), (6, 10)

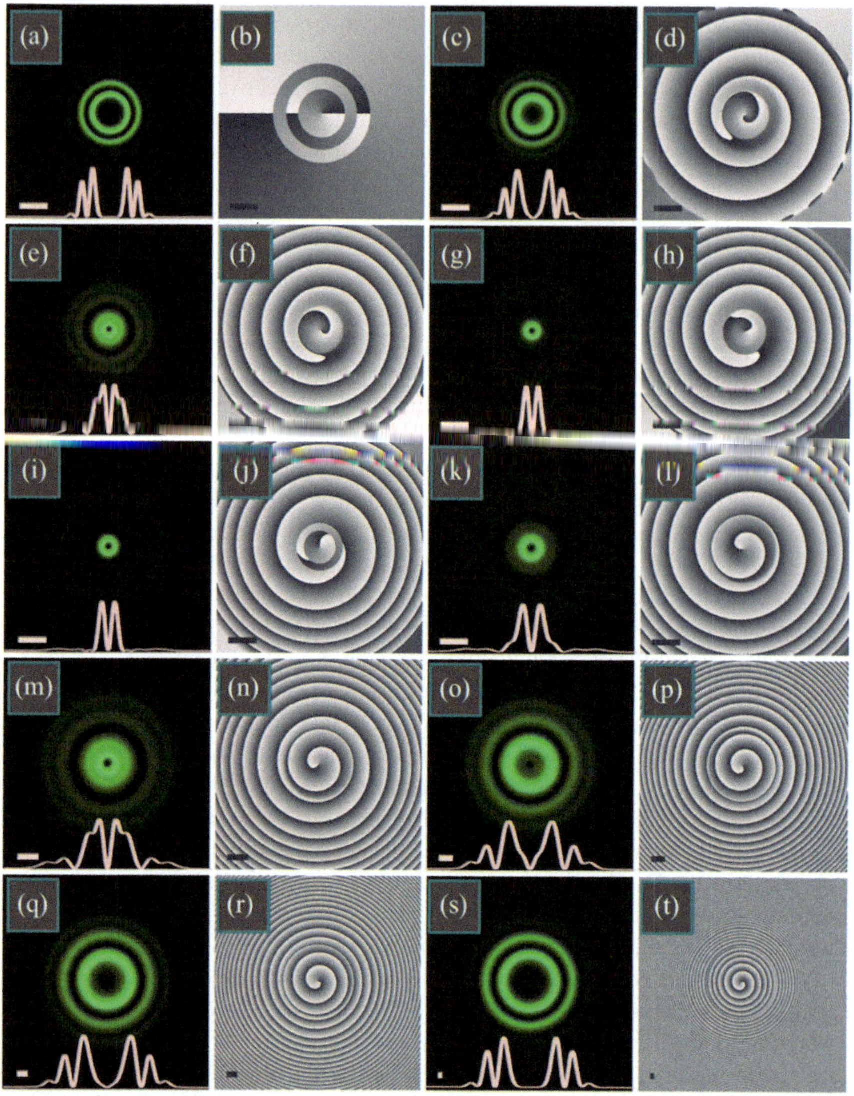

Fig. 3.8 Intensity (coumns 1 and 3) and phase distributions (columns 2 and 4) of the LG beam from Eq. (3.16) in different transverse planes $z/z_0 = 0$ (**a**, **b**), 1/4 (**c**, **d**), 1/2 (**e**, **f**), 3/4 (**g**, **h**), 1 (**i**, **j**), 1.5 (**k**, **l**), 2 (**m**, **n**), 3 (**o**, **p**), 4 (**q**, **r**), 10 (**s**, **t**). Other computation parameters are: wavelength $\lambda = 532$ nm, waist radius of the Gaussian beam $w = 0.5$ mm, azimuthal (upper) and radial (lower) indices of the associated Laguerre polynomial are respectively $m = 4$ and $n = 3$. Plots in the intensity distributions show the intensity cross-sections. Scale marks in left lower corners (light in columns 1 and 3 and dark in columns 2 and 4) show 1 mm

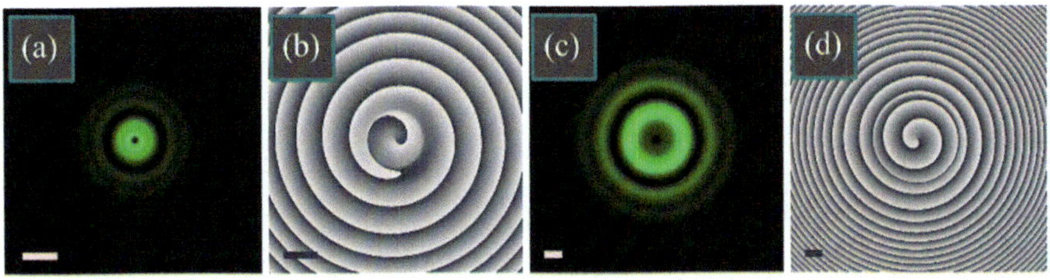

Fig. 3.9 The same intensity and phase distributions as in Fig. 3.8e, f, o, p, but obtained analytically by Eq. (3.21) with using the expansion coefficients (3.22)

(Fig. 3.10f). For these values, the topological charge of the beams in the left column of Fig. 3.8 is 1, whereas for the beams from the right column it is equal to 4. Since $n = 0$ in Fig. 3.10a, b, these beams reduce to the conventional single-ringed LG beams of the topological charge m. These beams do not possess the autofocusing property. At $n = 3$ and $n = 6$, Fig. 3.10 confirms that the beams are autofocusing, although the pronounced focus, when the intensity decays more than two times from its maximum, is only at $(n, m) = (6, 7)$. The depth-of-field in this case is almost equal to the Rayleigh length: DOF $\approx 0.9z_0$. In addition, Fig. 3.10 confirms that the radial index affect the size of the dark area inside the main light ring, but in the initial plane and in the far field. On the contrary, in the plane of the autofocus, the beam is more narrow that the conventional LG beam in its waist (the beam in Fig. 3.10e is more narrow than the beam in Fig. 3.10c, which is more narrow than the conventional LG beam from Fig. 3.10a). To demonstrate, to what extent we can increase or decrease the dark area in the focus, Fig. 3.11 depicts the beam (3.16) with a unit topological charge, but with a small and with a large index values: $(n, m) = (1, 2)$ (Fig. 3.11a–c) and $(n, m) = (20, 21)$ (Fig. 3.11d–f). At $(n, m) = (1, 2)$, the focal distance decreases to $0.7z_0$, and the ring diameter is nearly equal to the ring diameter of the conventional LG beam with the charge of 1 in its waist, i.e. 0.7 mm ($2^{1/2}w$). At $(n, m) = (20, 21)$, the beam is much narrower in the focal plane: its main ring has the diameter of nearly 0.2 mm. The depth-of-field in this case is nearly equal to the half of Rayleigh length: DOF $\approx 0.5z_0$. We believe that further increasing of the values n and m will lead to further decreasing (theoretically unlimited) of the ring radius in the autofocus plane, but in this case the light beam becomes nonparaxial and we cannot use Eq. (3.30) and the Fresnel transform to estimate the ring diameter in the focal plane.

Here, we investigated a new type of Laguerre-Gaussian beam with autofocusing [41]. At a fixed value of the topological charge l of these beams, changing the radial index n of the Laguerre polynomial allows controlling the focusing degree of these beams at the Rayleigh distance from the waist. Although the numerical aperture of such focusing is not high (inversely proportional to the Rayleigh distance $z0$), the depth of field, in this case, is significant (proportional to z0). Another interesting property of these beams, in contrast to the conventional Laguerre-Gaussian beams, is that at the zero topological charge (l = 0), the beams (3.20) generate in the focus of a spherical lens a light ring. The diameter of this ring can be adjusted by changing the radial index n and the ratio of the lens focal length and the Rayleigh distance (f/

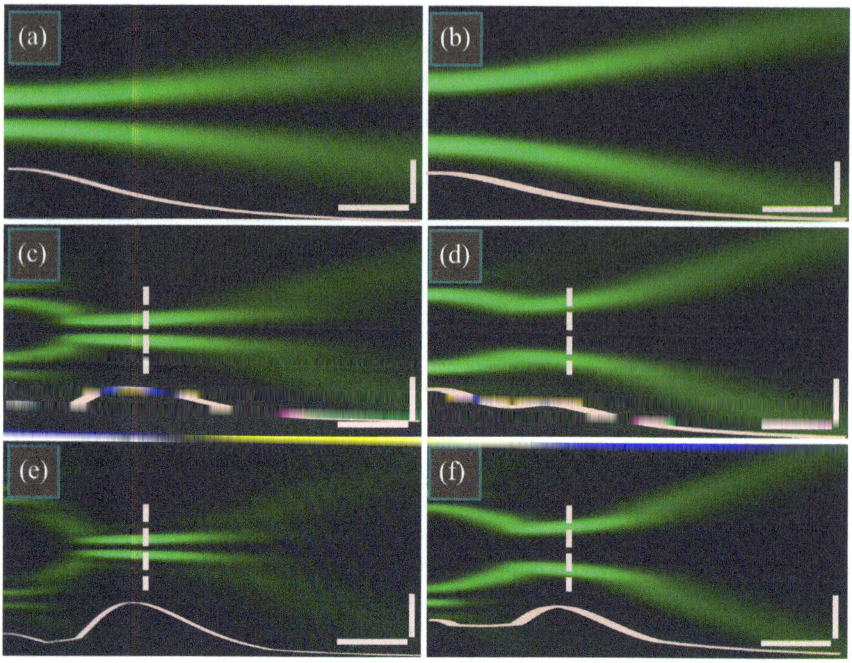

Fig. 3.10 Longitudinal intensity distributions of the beam from Eq. (3.16) with the topological charge of 1 (**a**, **c**, **e**) and 4 (**b**, **d**, **f**) when the radial (lower) index n of the associated Laguerre polynomial is as 0 (**a**, **b**), 3 (**c**, **d**), 6 (**e**, **f**). Below each distribution, longitudinal intensity cross-section is shown. Horizontal and vertical scale marks denote respectively $z_0/2$ and 1 mm. Transverse dashed line denotes the autofocus plane $z = z_0$

z0). This property of these beams can be used for simultaneous optical trapping of several metallic microparticles into the increased dark area of the light ring in the focus of the spherical lens. We note that the Bessel or Bessel-Gaussian [42] beams are also able to generate a light ring in focus at a zero topological charge. However, the Bessel and Bessel-Gaussian beams are not Fourier-invariant and generate in the focus of one light ring. The investigated here LG beams are Fourier-invariant and can generate in the focus several light rings with nearly the same intensity.

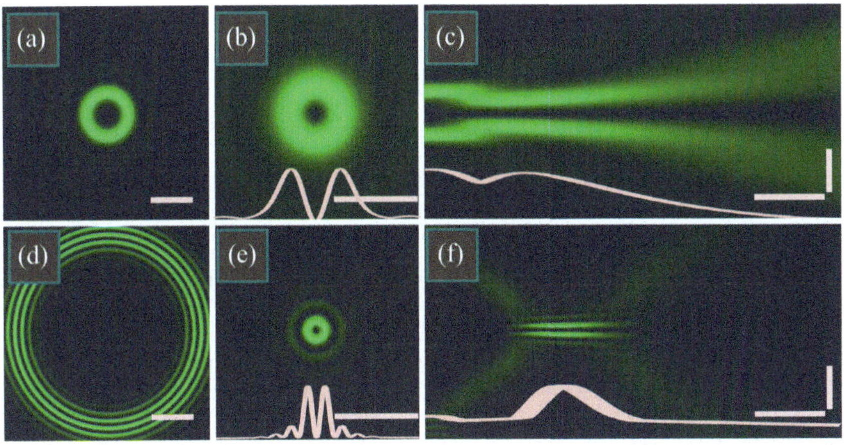

Fig. 3.11 Transverse initial (**a**, **d**) and focal (**b**, **e**) intensity distributions, as well as the longitudinal intensity distributions (**c**, **e**) of the beam from Eq. (3.1) with the $(n, m) = (1, 2)$ (**a**–**c**) and with the $(n, m) = (20, 21)$ (**d**, **f**). Curves (**b**, **c**, **e**, **f**) show the intensity cross-sections. Horizontal scale marks in the transverse patterns denote 1 mm. Horizontal and vertical scale marks in longitudinal pattern denote respectively $z_0/2$ and 1 mm

References

1. H. Kogelnik, T. Li, Laser beams and resonators. Appl. Opt. **5**, 1550–1567 (1966)
2. E. Zauderer, Complex argument Hermite-Gaussian and Laguerre-Gaussian beams. J. Opt. Soc. Am. A **3**, 465–469 (1986)
3. A. Wünsche, Generalized Gaussian beam solutions of paraxial optics and their connection to a hidden symmetry. J. Opt. Soc. Am. A **6**, 1320–1329 (1989)
4. E. Abramochkin, V. Volostnikov, Beam transformations and nontransformed beams. Opt. Commun.Commun. **83**, 123–135 (1991)
5. L. Allen, M.W. Beijersbergen, R.J.C. Spreeuw, J.P. Woerdman, Orbital angular momentum of light and the transformation of Laguerre-Gaussian laser modes. Phys. Rev. A **45**, 8185–8189 (1992)
6. E.G. Abramochkin, V.G. Volostnikov, Generalized Gaussian beams. J. Opt. A Pure Appl. Opt. **6**, S157–S161 (2004)
7. E. Abramochkin, E. Razueva, V. Volostnikov, General astigmatic transform of Hermite–Laguerre-Gaussian beams. J. Opt. Soc. Am. A **27**, 2506–2513 (2010)
8. G. Zhou, G. Ru, Orbital angular momentum density of an elegant Laguerre-Gaussian beam. Prog. Electromagnet. Res. **141**, 751–768 (2013)
9. V.V. Kotlyar, S.N. Khonina, A.A. Almazov, V.A. Soifer, K. Jefimovs, J. Turunen, Elliptic Laguerre-Gaussian beams. J. Opt. Soc. Am. A **23**, 43–56 (2006)
10. J. Mendoza-Hernández, M.L. Arroyo-Carrasco, M.D. Iturbe-Castillo, S. Chávez-Cerda, Laguerre-Gauss beams versus Bessel beams showdown: peer comparison. Opt. Lett. **40**, 3739–3742 (2015)
11. M. Wang, Y. Ma, Q. Sheng, X. He, J. Liu, W. Shi, J. Yao, T. Omatsu, Laguerre-Gaussian beam generation via enhanced intracavity spherical aberration. Opt. Express **29**, 27783–27790 (2021)
12. M. Rafayelyan, E. Brasselet, Laguerre-Gaussian modal q-plates. Opt. Lett. **42**, 1966–1969 (2017)

13. H. Mao, Y.H. Ren, Y. Yu, Z. Yu, X. Sun, S. Zhang, K.K. Wong, Broadband meta-converters for multiple Laguerre-Gaussian modes. Photon. Res. **9**, 1689–1698 (2021)
14. G. Liang, Q. Wang, Controllable conversion between Hermite Gaussian and Laguerre Gaussian modes due to cross phase. Opt. Express **27**, 10684–10691 (2019)
15. A. Longman, R. Fedosejevs, Optimal Laguerre-Gaussian modes for high-intensity optical vortices. J. Opt. Soc. Am. A **37**, 841–848 (2020)
16. M. Dong, X.Y. Lu, C. Zhao, Y. Cai, Y. Yang, Measuring topological charge of partially coherent elegant Laguerre-Gaussian beam. Opt. Express **26**, 33035–33043 (2018)
17. A.A. Kovalev, V.V. Kotlyar, A.P. Porfirev, Asymmetric Laguerre-Gaussian beams. Phys. Rev. A **93**, 063858 (2016)
18. Y.H. Hsieh, Y.H. Lai, M.X. Hsieh, K.F. Huang, Y.F. Chen, Generating high-power asymmetrical Laguerre-Gaussian modes and exploring topological charges distribution. Opt. Express **26**, 31738–31749 (2018)
19. M. Ghaderi Goran Abad, M. Mahmoudi, Laguerre-Gaussian modes generated vector beam via nonlinear magneto-optical rotation. Sci. Rep. **11**, 5972 (2021)
20. S. Huang, Z. Miao, C. He, F. Pang, Y. Li, T. Wang, Composite vortex beams by coaxial superposition of Laguerre-Gaussian beams. Opt. Lasers Eng. **78**, 132–139 (2016)
21. Y. Chen, L. Liu, F. Wang, C. Zhao, Y. Cai, Elliptical Laguerre-Gaussian correlated Schell-model beam. Opt. Express **22**, 13975–13987 (2014)
22. K. Pang, C. Liu, G. Xie, Y. Ren, Z. Zhao, R. Zhang, Y. Cao, J. Zhao, H. Song, L. Li, A.N. Willner, M. Tur, R.W. Boyd, A.E. Willner, Demonstration of a 10 Mbit/s quantum communication link by encoding data on two Laguerre-Gaussian modes with different radial indices. Opt. Lett. **43**, 5639–5642 (2018)
23. T. Doster, A.T. Watnik, Laguerre-Gauss and Bessel-Gauss beams propagation through turbulence: analysis of channel efficiency. Appl. Opt. **55**, 10239–10246 (2016)
24. J. Zeng, X. Liu, C. Zhao, F. Wang, G. Gbur, Y. Cai, Y. Spiral spectrum of a Laguerre-Gaussian beam propagating in anisotropic non-Kolmogorov turbulent atmosphere along horizontal path. Opt. Express **27**, 25342–25356 (2019)
25. M.A. Cox, L. Maqondo, R. Kara, G. Milione, L. Cheng, A. Forbes, The resilience of Hermite- and Laguerre-Gaussian modes in turbulence. J. Lightwave Technol. **37**, 3911–3917 (2019)
26. T. Otsu, T. Ando, Y. Takiguchi, Y. Ohtake, H. Toyoda, H. Itoh, Direct evidence for three-dimensional off-axis trapping with single Laguerre-Gaussian beam. Sci. Rep. **4**, 4579 (2014)
27. A.A. Peshkov, D. Seipt, A. Surzhykov, S. Fritzsche, Photoexcitation of atoms by Laguerre-Gaussian beams. Phys. Rev. A **96**, 023407 (2017)
28. A.P. Prudnikov, Y.A. Bychkov, O.I. Marychev, *Integrals and Series: Volume 2: Special Functions* (Nauka Publisher, Moscow, 1983)
29. V.V. Kotlyar, A.A. Kovalev, E.S. Kozlova, A.A. Savelyeva, S.S. Stafeev, New type of vortex laser beams: Squared Laguerre-Gaussian beam. Optik **270**, 169916 (2022)
30. V.V. Kotlyar, E.G. Abramochkin, A.A. Kovalev, A.A. Savelyeva, Product of two Laguerre-Gaussian beams. Photonics. **9**, 496 (2022)
31. P.A. Prentice, M.P. MacDonald, T.G. Frank, A. Cuschieri, G.C. Spalding, W. Sibbett, P.A. Campbell, K. Dholakia, Manipulation and filtration of low index particles with holographic Laguerre-Gaussian optical trap arrays. Opt. Express **12**, 593–600 (2004)
32. N.A. Ferlic, M. van Iersel, D.A. Paulson, C.C. Davis, Propagation of Laguerre-Gaussian and I_m Bessel beams through atmospheric turbulence: a computational study. Proc. SPIE **11506**, 115060H (2020)
33. M. Cao, Y. Yu, L. Zhang, F. Ye, Y. Wang, D. Wei, P. Zhang, W. Guo, S. Zhang, H. Gao, F. Li, Demonstration of CNOT gate with Laguerre Gaussian beams via four-wave mixing in atom vapor. Opt. Express **22**, 20177–20184 (2014)
34. P. Dedecker, B. Muls, J. Hofkens, J. Enderlein, J. Hotta, Orientational effects in the excitation and de-excitation of single molecules interacting with donut-mode laser beams. Opt. Express **15**, 3372–3383 (2007)
35. N. Bokor, Y. Iketaki, T. Watanabe, M. Fujii, Investigation of polarization effects for high-numerical-aperture first-order Laguerre-Gaussian beams by 2D scanning with a single fluorescent microbead. Opt. Express **13**, 10440–10447 (2005)

36. G. Zhou, G. Ru, Orbital angular momentum density of an elegant Laguerre-Gaussian beam. Prog. Electromagn. Res. **141**, 751–768 (2013)
37. J.F. Bisson, Y. Senatsky, K.I. Ueda, Generation of Laguerre-Gaussian modes in Nd:YAG laser using diffractive optical pumping. Laser Phys. Lett. **2**, 327–333 (2005)
38. D. Lin, J.M.O. Daniel, W.A. Clarkson, Controlling the handedness of directly excited Laguerre-Gaussian modes in a solid-state laser. Opt. Lett. **39**, 3903–3906 (2014)
39. M.P. Thirugnanasambandam, Y. Senatsky, K. Ueda, Generation of very-high order Laguerre-Gaussian modes in Yb:YAG ceramic laser. Laser Phys. Lett. Phys. Lett. **7**, 637–643 (2010)
40. N. Matsumoto, T. Ando, T. Inoue, Y. Ohtake, N. Fukuchi, T. Hara, Generation of high-quality higher-order Laguerre-Gaussian beams using liquid-crystal-on-silicon spatial light modulators. J. Opt. Soc. Am. A **25**, 1642–1651 (2008)
41. V.V. Kotlyar, E.G. Abramochkin, A.A. Kovalev, A.A. Savelyeva, Laguerre-Gaussian beams with an increased dark area and autofocusing. Photonics **9**, 708 (2022)
42. F. Gori, G. Guattari, C. Padovani, Bessel-Gauss beams. Opt. Commun.Commun. **64**, 491–495 (1987)

Chapter 4
New Type of Bessel-Gaussian Beams

4.1 Double and Square Bessel–Gaussian Beams

Starting from the well-known work by Gori et al. [1], miscellaneous modifications of the Bessel-Gaussian (BG) beams are investigated with a growing interest. Bessel-Gaussian beams are convenient for probing the turbulent atmosphere, since, on the one hand, they are of finite-energy like the Gaussian beam, and, on the other hand, they manifest quasi diffraction-free properties, due to the Bessel function [2–5]. As was shown in [6], the BG beams are more resistant to the distortions induced by turbulent atmosphere, than the Gaussian beam. The work [7] demonstrated that in the Kolmogorov-type turbulent atmosphere, the BG beams conserve along a longer distance, compared to the Gaussian beam. In [8], modified BG beams were studied, that were called "frozen waves". Free-space propagation of the BG beams in THz wavelengths range was studied in [9]. In [10], propagation of asymmetric BG beams [11, 12] in the turbulent atmosphere was considered. Bessel-Gaussian beams are used for microparticles manipulation [13, 14] and for generating entangled pairs of photons in quantum informatics [15, 16]. The BG beams can be generated by an axicon, by a light modulator or by other elements [17–19]. Therefore, extending the set of different types and modifications of the BG beams is a relevant problem.

In this section, we obtain modifications of the BG beams, such as square BG beams and products of the BG beams. For such beams, we derive explicit analytical expressions that describe their evolution on propagation in free space. Similarly to standard BG beams, the beams we studied are not structurally stable, but they are also composed of a set of concentric light rings. The number of the rings and light energy between them are changing upon the beam propagation in free space. In the far field, all the beams studied except one are similar to the standard BG beam and contain a single light ring with almost no side lobes. Only one beam, vortex-free power-function BG beam, contains several rings in the far field and their number equals to the beam order.

V. V. Kotlyar et al., *Helical Laser Beams*, https://doi.org/10.1007/978-3-031-92217-6_4

4.1.1 Bessel-Gaussian Beams and Modulated Bessel-Gaussian Beams

Bessel-Gaussian (BG) beams, first studied in [1], are not structurally stable ones, i.e. their transverse intensity distribution changes when the beams propagate in free space. However, the change is insignificant and all the intensities keep their shape of concentric circles. The complex amplitude of the BG beam at an arbitrary distance along the optical axis z is

$$\mathrm{BG}_m(r, \varphi, z|c) = \frac{1}{q(z)} \exp\left(\frac{c^2}{4q(z)} - \frac{r^2}{w_0^2 q(z)} + im\varphi\right) J_m\left(\frac{cr}{w_0 q(z)}\right), \tag{4.1}$$

where $q(z) = 1 + iz/z_0$, $z_0 = kw_0^2/2$ is the Rayleigh range, (r, φ) are the polar coordinates in the transverse plane, m is the topological charge, $J_m(x)$ is the mth-order Bessel function of the first kind, w_0 is the Gaussian beam waist radius, k is the wavenumber of light, c is a dimensionless (possibly complex) parameter, affecting the transverse component of the wave vector ($c/w_0 = k_r$), $k^2 = k_r^2 + k_z^2$, k_z is the longitudinal component of the wave vector. Here and in the following text, we will use a vertical line to separate variables and parameters.

Complex amplitude of the BG beam (4.1) in the initial plane can be expanded into a series of Laguerre-Gaussian (LG) beams:

$$\begin{aligned}\mathrm{BG}_m(r, \varphi, 0|c) &= \exp\left(\frac{c^2}{4} - \frac{r^2}{w_0^2} + im\varphi\right) J_m\left(\frac{cr}{w_0}\right) \\ &= \exp\left(\frac{c^2}{8}\right) \sum_{\nu=0}^{\infty} \frac{c^{2\nu+m} \mathrm{LG}_{\nu,m}(r, \varphi)}{2^{3\nu+m}(\nu+m)!},\end{aligned} \tag{4.2}$$

where LG beams in the initial plane are described by the following expression:

$$\mathrm{LG}_{n,m}(r, \varphi) = \exp\left(-\frac{r^2}{w_0^2} + im\varphi\right)\left(\frac{r}{w_0}\right)^m L_n^m\left(\frac{2r^2}{w_0^2}\right), \tag{4.3}$$

based on the associated Laguerre polynomial $L_n^m(x)$. As seen from Eq. (4.2), when the parameter c is small ($c \ll 1$), only the first term in the series is significant and the BG beam in this case is close to the LG beam.

In addition to standard BG beams, BG beams with quadratic radial dependence (qBG) are also known. They were introduced in [20]. Later, in [21] the transformation of these beams by an optical ABCD system was investigated, and now qBG beams are considered as a part of the classification of circular optical beams [22]. In the initial plane, the complex amplitude of qBG beams is defined as

$$\mathrm{qBG}_m(r, \varphi, 0|c) = \exp\left(-\frac{r^2}{w_0^2} + im\varphi\right) J_{m/2}\left(\frac{cr^2}{w_0^2}\right). \tag{4.4}$$

As seen from Eq. (4.4), the qBG beams depends on the Bessel function of an integer order when m is even and of a half-integer order when m is odd. It is also seen that, in contrast with BG beams, the argument of the Bessel function depends on the radial coordinate quadratically. The complex amplitude of the qBG beams in the Fresnel zone is as follows:

$$\mathrm{qBG}_m(r,\varphi,z|c) = \frac{1}{\sqrt{q_+q_-}}\exp\left(-\left[1+i(1+c^2)\frac{z}{z_0}\right]\frac{r^2}{w_0^2q_+q_-}+im\varphi\right)J_{m/2}\left(\frac{cr^2}{w_0^2q_+q_-}\right). \quad (4.5)$$

The parameters $q_\pm$ in Eq. (4.4) depend on the propagation distance z and the parameter c:

$$q_\pm = q_\pm(z) = q(z) \pm c\frac{z}{z_0} = 1+(i\pm c)\frac{z}{z_0}. \quad (4.6)$$

Equations (4.5) and (4.6) reveal that when the value c is large enough ($c \gg 1$), the imaginary part of the factor q_+q_- in the argument of the Bessel and the exponential functions is small compared to the real part, and the BG becomes approximately propagation-invariant, i.e. its shape of the intensity distribution is almost conserved, changing only in scale.

Further, we obtain an integral transform that relates BG beams (4.1) with qBG beams (4.4). It can be shown that this transform is given by

$$\int_0^\infty \mathrm{BG}_m(r,\varphi,z|c)J_{m/2}\left(\frac{ac^2}{4}\right)c\,dc = \frac{2}{\sqrt{1+a^2q^2(z)}}\exp\left(-\frac{a^2q(z)r^2}{1+a^2q^2(z)}+im\varphi\right)J_{m/2}\left(\frac{ar^2}{1+a^2q^2(z)}\right). \quad (4.7)$$

Comparing right parts of Eqs. (4.5) and (4.7) reveals that the complex amplitudes depend on the parameters a and c differently, although the structure of the modulated BG beam is the same in Eqs. (4.5) and (4.7). Пучки (4.7) являются модификацией пучков (4.4), (4.5).

4.1.2 Square Bessel-Gaussian Beams

In our works [23, 24], we considered double Laguerre-Gaussian (dLG) beams and square LG beams. Both of them can be expressed via finite sums of LG beams. Here we consider square BG beams and show that they can be represented as an infinite sum of BG beams (4.1). Complex amplitude of the square BG beam (sBG) in the initial plane reads as

$$\mathrm{sBG}_m(r,\varphi,0|c)=\exp\left(\frac{c^2}{2}-\frac{r^2}{w_0^2}+2im\varphi\right)J_m^2\left(\frac{cr}{w_0}\right). \tag{4.8}$$

Here we prefer to scale the parameters of the initial BG beam ($c \to c\sqrt{2}$, $w_0 \to w_0\sqrt{2}$) but to keep the Gaussian factor unchanged.

The complex amplitude of the BBG beams can be obtained by using generating function for the squares of the Bessel functions [25]:

$$\sum_{\nu\in\mathbb{Z}} J_\nu^2(x)(-t^2)^\nu = J_0\left[x\left(t+\frac{1}{t}\right)\right]. \tag{4.9}$$

After rather complicated algebra, we reduce the Fresnel transform of the initial complex amplitude (4.8) to the following series of Bessel functions:

$$\begin{aligned}\mathrm{sBG}_m(r,\varphi,z|c) &= \frac{1}{q(z)}\exp\left(\frac{c^2}{2q(z)}-\frac{r^2}{w_0^2q(z)}+2im\varphi\right)\\ &\times\sum_{\nu\in\mathbb{Z}}(-i)^\nu J_{m-\nu}\left(\frac{cr}{w_0q(z)}\right)J_{m+\nu}\left(\frac{cr}{w_0q(z)}\right)J_\nu\left(\frac{c^2z}{2z_0q(z)}\right).\end{aligned} \tag{4.10}$$

Equation (4.10) indicates that the square BG beams do not conserve their shape on propagation in free space, but are a superposition of a countable number of products of BG beams of the orders, whose sum is equal to 2 m, i.e. the initial topological charge. It is interesting to note that in the far field, i.e. when $z \to \infty$, the series in Eq. (4.10) reduces to the power function r^{2m}. Thus, in the far field (and in the focus of a spherical lens), the square BG beam has the shape of a single light ring without side lobes.

4.1.3 Product of Two Bessel-Gaussian Beams

Since the series in Eq. (4.10) contains the product of two similar Bessel functions depending on r, it seems quite possible that for a product of two different BG beams instead of the squared one we can evaluate the Fresnel transform. Let us introduce the product of two BG beams in the initial plane $z = 0$:

$$\begin{aligned}\mathrm{dBG}_{m,n}(r,\varphi,0|a,b) &= \exp\left(\frac{a^2+b^2}{4}-\frac{r^2}{w_0^2}+i(m+n)\varphi\right)\\ &\times J_m\left(\frac{ar}{w_0}\right)J_n\left(\frac{br}{w_0}\right).\end{aligned} \tag{4.11}$$

Then in same way as the expansion (4.10) has been derived, the initial field (4.11) leads to the following solution of the paraxial equation:

$$\mathrm{dBG}_{m,n}(r,\varphi,z|a,b)=\frac{1}{q(z)}\exp\left(\frac{a^2+b^2}{4q(z)}-\frac{r^2}{w_0^2q(z)}+i(m+n)\varphi\right)\\ \times\sum_{\nu\in\mathbb{Z}}(-i)^{\nu}J_{m+\nu}\left(\frac{ar}{w_0q(z)}\right)J_{n-\nu}\left(\frac{br}{w_0q(z)}\right)J_{\nu}\left(\frac{abz}{2z_0q(z)}\right). \tag{4.12}$$

As can be seen from Eq. (4.12), if $n=m$ and $a=b=c$, the product of two BG beams reduces to the square BG beam. It is also seen that, for the case $n=b=0$, the series (4.12) collapses to the only term with $\nu=0$ and the product of two BG beams reduces to the standard BG beam:

$$\mathrm{dBG}_{m,0}(r,\varphi,z|a,0)=\mathrm{BG}_m(r,\varphi,z|a). \tag{4.13}$$

Some other particular cases of the dBG beams, Eq. (4.12), are also interesting to mention. If $n=-m$, then dBG beam is a vortex-free beam:

$$\mathrm{dBG}_{m,-m}(r,\varphi,0|a,b)=(-1)^m\exp\left(\frac{a^2+b^2}{4}-\frac{r^2}{w_0^2}\right)J_m\left(\frac{ar}{w_0}\right)J_m\left(\frac{br}{w_0}\right),\\ \mathrm{dBG}_{m,-m}(r,\varphi,z|a,b)=\frac{(-1)^m}{q(z)}\exp\left(\frac{a^2+b^2}{4q(z)}-\frac{r^2}{w_0^2q(z)}\right)\\ \times\sum_{\nu\in\mathbb{Z}}i^{\nu}J_{m+\nu}\left(\frac{ar}{w_0q(z)}\right)J_{m+\nu}\left(\frac{br}{w_0q(z)}\right)J_{\nu}\left(\frac{abz}{2z_0q(z)}\right). \tag{4.14}$$

Its limiting case, when b vanishes, is a product of the vortex-free BG beam by the power function:

$$\mathrm{pBG}_m(r,\varphi,0|a)=2^m m!\cdot\lim_{b\to 0}\frac{\mathrm{BBG}_{m,-m}(r,\varphi,0|a,b)}{(-ab)^m}\\ =\exp\left(\frac{a^2}{4}-\frac{r^2}{w_0^2}\right)J_m\left(\frac{ar}{w_0}\right)\left(\frac{2r}{aw_0}\right)^m. \tag{4.15}$$

The beam from Eq. (4.15) can be called vortex-free power-function BG beam. Tending the parameter b to zero in Eq. (4.12), we derive the Fresnel transform of the beam from Eq. (4.15):

$$\mathrm{pBG}_m(r,\varphi,z|a)=\frac{1}{q(z)}\exp\left(\frac{a^2}{4q(z)}-\frac{r^2}{w_0^2q(z)}\right)$$

$$\times \sum_{\nu=0}^{m} \binom{m}{\nu} \left(\frac{2r}{aw_0 q(z)} \right)^{m-\nu} J_{m-\nu}\left(\frac{ar}{w_0 q(z)} \right) \left(\frac{iz}{z_0 q(z)} \right)^{\nu}. \tag{4.16}$$

Expression (4.16) indicates that the power-function vortex-free BG beam in the Fresnel diffraction zone is a finite superposition of similar power-function BG beams of the orders ν from 0 to m.

In the far field ($z \gg z_0$), the argument of the Bessel functions in Eq. (4.16) becomes small. Thus, since $J_\nu(\xi) \approx (\xi/2)^\nu / \nu!$ at $\xi \approx 0$, the field from Eq. (4.16) transforms into a superposition of single-ringed beams:

$$\mathrm{pBG}_m(r, \varphi, z \gg z_0 | u) \approx \frac{1}{q(z)} \exp\left(\frac{a^2}{4q(z)} - \frac{r^2}{w_0^2 q(z)} \right) \times \sum_{\nu=0}^{m} (-1)^\nu \binom{m}{\nu} \frac{1}{\nu!} \left(\frac{z_0}{z} \frac{r}{w_0} \right)^{2\nu} \tag{4.17}$$

Each if these beams has its own light ring and, consequently, there should be m light rings in the far field.

4.1.4 Numerical Simulation

In this section, we describe the computation results of the beams from Eqs. (4.5), (4.12), and (4.16). All distributions are obtained by two ways: by using the numerical Fresnel transform, implemented as a convolution with adopting the fast Fourier transform, and by the theoretical expression. All intensity distributions, obtained by these two ways, are visually indistinguishable, while the phase distributions are different only in low-intensity areas. This confirms the correctness of the formulae (4.5), (4.12), (4.16) for the complex amplitudes on space propagation.

Shown in Fig. 4.1 are intensity and phase distributions of the modulated Bessel-Gaussian beam (4.5) in several transverse planes. To obtain a beam with several rings and to prove that it is approximately propagation-invariant, we choose a large value of the scaling factor $c = 30$.

Figure 4.1 confirms that when the scaling factor c is large enough, the transverse intensity shape almost does not change upon propagation.

Figure 4.2 illustrates the intensity and phase distributions of two BG beams with different parameters, as well as of the beam (4.11), (4.12), constructed as their product, in several transverse planes. For computation on the base of Eq. (4.12), the series was bounded by 100 terms.

As seen in Fig. 4.2, intensity distributions of both BG beams in the initial plane have a shape of one bright ring (Fig. 4.2a, c), but the first beam has the ring of a smaller radius and also has a pale second ring (Fig. 4.2a). Between these rings, there is a dark zero-intensity ring and multiplication of the complex amplitudes of both

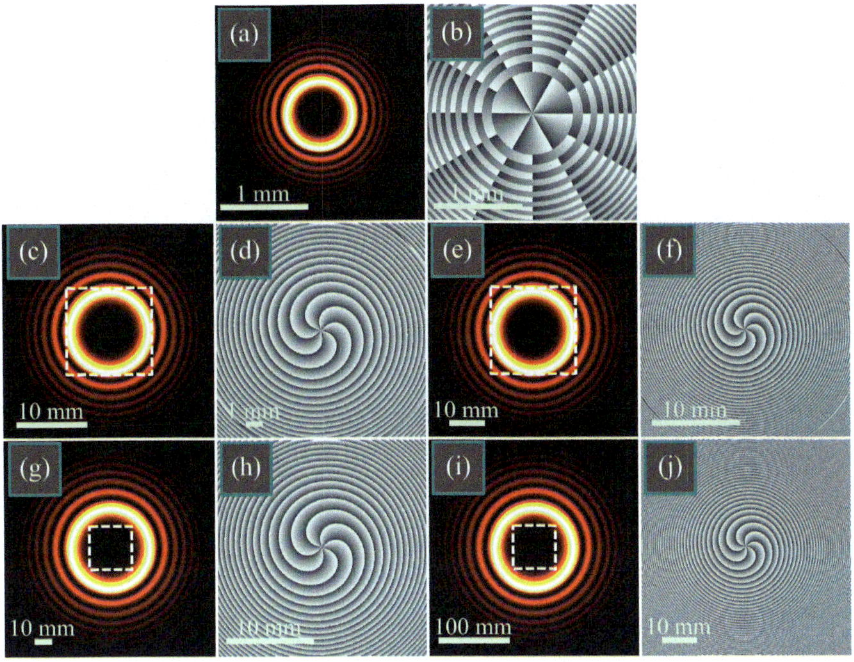

Fig. 4.1 Intensity (**a**, **c**, **e**, **g**, **i**) and phase (**b**, **d**, **f**, **h**, **j**) distributions of the modulated Bessel-Gaussian beam (4.5) in several transverse planes for the following computation parameters: wavelength λ = 532 nm, Gaussian beam waist radius w_0 = 1 mm, beam order $m = 3$, scaling factor $c = 30$, propagation distances $z = 0$ (**a**, **b**), $z = z_0/2$ (**c**, **d**), $z = z_0$ (**e**, **f**), $z = 2z_0$ (**g**, **h**), $z = 5z_0$ (**i**, **j**). Dashed squares (**c**, **e**, **g**, **i**) denote the areas corresponding to shown phase distributions (**d**, **f**, **h**, **j**)

beams in the beam (4.11), (4.12) leads to two bright light rings (Fig. 4.2e), since the thick ring in Fig. 4.2c is "cut" into two rings by the dark ring from Fig. 4.2a.

In the initial phase distributions (Fig. 4.2b, d, f), there are rings with the phase jumps by π. The Bessel functions are equal to zero on these rings. However, on propagation, the arguments of the Bessel functions become complex and the function values are nonzero. Therefore, there are no such phase jumps in Fig. 4.2h, j, l, n, p, r, t, v, x.

Despite the lower topological charge of the first BG beam ($m = 2$ vs. $n = 3$), its scaling factor is, vice versa, greater than that of the second beam ($a = 8$ vs. $b = 5$). Therefore, on propagation, it diverges faster and at a distance of $z = z_0/2$ its light ring has a greater diameter than the ring of the second BG beam, and now it is this ring that becomes cut by the minimal-intensity ring of the second beam. Therefore, the beam in Fig. 4.2k, l has also two light rings, as in the initial plane in Fig. 4.2e, f. On propagation to the Fresnel zone and to the far field (row 3 and 4 in Fig. 4.2), the light rings of both BG beams almost do not overlap (rings in Fig. 4.2m, o in the Fresnel zone and rings in Fig. 4.2s, u in the far field). Therefore, after multiplication, these rings are suppressed and another rings appear, those that are not seen in the initial

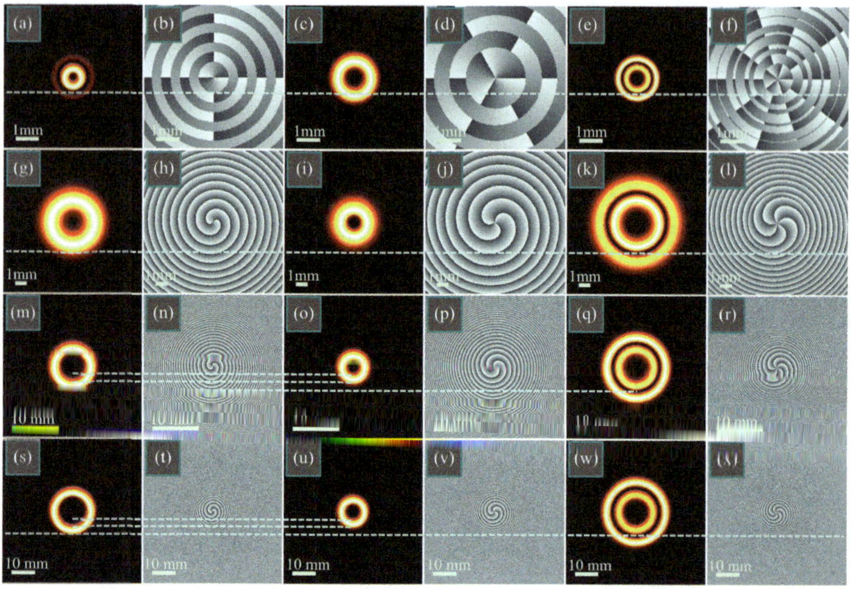

Fig. 4.2 Intensity (columns 1, 3, 5) and phase (columns 2, 4, 6) distributions of two BG beams with different parameters (columns 1–4), as well as of the beam (4.11), (4.12), constructed as their product (columns 5, 6), in several transverse planes for the following computation parameters: wavelength $\lambda = 532$ nm, Gaussian beam waist radius $w_0 = 1$ mm, orders of BG beams $m = 2$ (columns 1, 2) and $n = 3$ (columns 3, 4), scaling factors of BG beams $a = 8$ (columns 1, 2) and $b = 5$ (columns 3, 4), propagation distances $z = 0$ (row 1), $z = z_0/2$ (row 2), $z = z_0$ (row 3), $z = 2z_0$ (row 4). Horizontal dashed lines between the figures are shown for matching the radii of the light rings in different beams

plane. Thus, the diameter of the outer light ring of the product beam (Fig. 4.2q, w) exceeds the ring diameters of both BG beams.

Figure 4.3 depicts the intensity and phase distributions of the vortex-free power-function BG beam (4.16) in several transverse planes.

As seen in Fig. 4.3, the intensity distribution in the initial plane consists of multiple light rings (there are 7 rings in Fig. 4.3a). On propagation, only two rings remain, and then, in the far field, the number of rings increases to three, which is consistent with the theory that predicts that there should be m rings in the far field. The phase distribution is not shown for the initial plane since it is zero, whereas in other planes it is seen to be a rotationally symmetric and the beam does not contain optical vortices.

In this section, we have obtained the following results [26]. An integral transform (4.7) has been derived that relates the standard BG beams (4.1), (4.3) and модифицированные qBG beams, отличающиеся от стандартных qBG beams (4.4), (4.5). As seen from Eq. (4.7), the beams can be treated as continuous superposition of the standard BG beams with the weight function equal to fractional-order Bessel function. Square BG (BBG) beams (4.8) have been proposed and studied, whose complex amplitude depends on the square of the Bessel function. We have

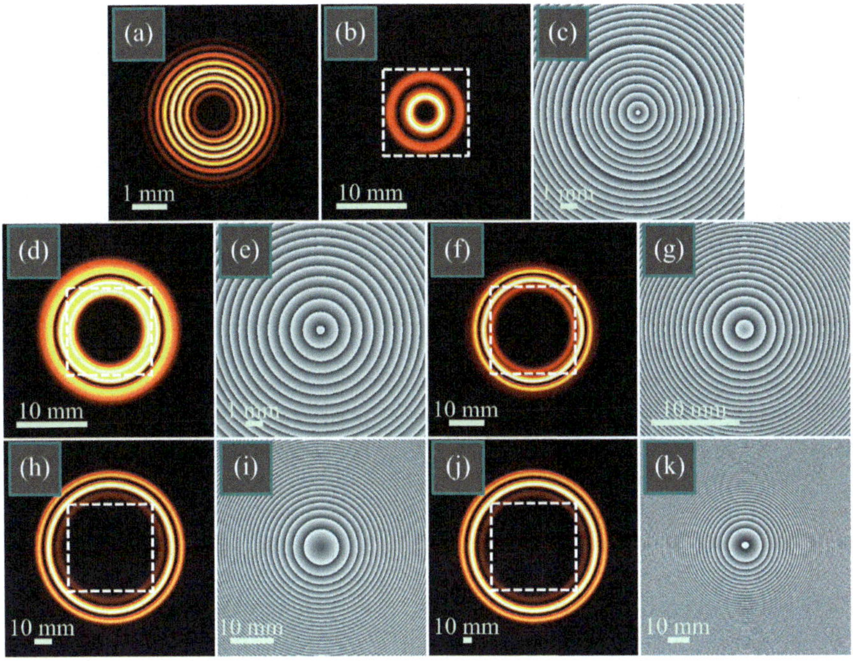

Fig. 4.3 Intensity (**a**, **b**, **d**, **f**, **h**, **j**) and phase (**c**, **e**, **g**, **i**, **k**) distributions of the vortex-free power-function Bessel-Gaussian beam (4.16) in several transverse planes for the following computation parameters: wavelength $\lambda = 532$ nm, Gaussian beam waist radius $w_0 = 1$ mm, beam order $m = 3$, scaling factor $a = 15$, propagation distances are $z = 0$ (**a**), $z = z_0/2$ (**b**, **c**), $z = z_0$ (**d**, **e**), $z = 2z_0$ (**f**, **g**), $z = 5z_0$ (**h**, **i**), $z = 10z_0$ (**j**, **k**). Dashed squares (**b**, **d**, **f**, **h**, **j**) denote the areas corresponding to shown phase distributions (**c**, **e**, **g**, **i**, **k**)

obtained the complex amplitude of the BBG beams in the Fresnel diffraction zone in a form of a series of products of three different Bessel functions (4.10). As a generalization of square BG beams, we have also investigated double BG (dBG) beams (4.11) with their complex amplitude being proportional to a product of two Bessel functions of different orders and of different scales. Complex amplitude of such beams in the Fresnel diffraction zone has also been represented in a form of a series of products of three different Bessel functions (4.12). We have also considered modified BG beams with their complex amplitude being equal to a product of the Bessel function by the power function of the radial variable (4.15). This set of pBG beams is a subset of vortex-free BG beams. When such a beam of m-th order propagates in free space, it becomes a superposition of a finite number of similar vortex-free power-function BG beams of the orders from 0 to m. New varieties of the BG beams considered in this section will be useful for probing the atmosphere, wireless communications, microparticles manipulations, and in quantum informatics for generating entangled pairs of photons. In micromechanics, these laser beams can be used to control the movement of microparticles along circular trajectories [27–29].

4.2 Bessel-Bessel-Gaussian Vortex Laser Beams

The Bessel [30] and the Bessel-Gaussian [1] beams are known in optics since 1987. The work [31] contains a contemporary review of the Bessel beams. These beams still receive much attention due to their wide applications for nanostructuring materials [32–34], manipulating micropartcles [35–37], accelerating atoms [38–40], microscopy [41–43], and wireless optical communications [44–46]. Among various Bessel beams, many their modifications are known. One such a modification is a Bessel-Gaussian beam of a half-integer order and with an argument, quadratically dependent on the radial polar coordinate [20]. Another Bessel beam has also a half-integer order, quadratic argument, and finite energy, but it does not have the Gaussian envelope [47]. Asymmetric Bessel-Gaussian beams [12] are similar to the beams we study in this work, but the intensity shape of the former consists of one 'crescent' rather than of two 'crescents'. In Refs. [48] and [49], Bessel-Gaussian beams are compared in details with Laguerre-Gaussian beams and generalized (dual) Laguerre-Gaussian beams, respectively. In Ref. [26], double and square Bessel-Gaussian beams were introduced and investigated.

Besides employing a narrow ring aperture [30], which is, surely, very ineffective, Bessel beams can be generated by a deformable mirror, spatial light modulator, micromirrors array, binary mask [50], hologram [51], and by using an axicon with a lens (or a second axicon) for generating a bright ring [52–54], or by a toroidal cavity [55].

In this section, we obtain and investigate a new one-parametric family of Bessel-Bessel-Gaussian beams (BBG beams). We derive their complex amplitude as a product of the Gaussian function and two nth-order Bessel functions with a similar dependence of their arguments on the cylindrical coordinates. It is shown that changing the positive beam parameter allows controlling the beam topological charge. We also demonstrate anomalously fast transverse rotation of the beam intensity pattern upon propagation in the near field of diffraction. At a distance from the waist much shorter than the Rayleigh length, the intensity is already rotated by almost 45°.

4.2.1 Bessel-Bessel-Gaussian Beams Based Upon the Bessel-Gaussian Beams

Complex amplitude of the well-known Bessel-Gaussian beams [1] is given by

$$BG_n(r,\varphi,z)=\frac{1}{q}\exp\left(-\frac{i\alpha^2 z}{2kq}-\frac{r^2}{w^2 q}+in\varphi\right)J_n\left(\frac{\alpha r}{q}\right), \tag{4.18}$$

where $q = q(z) = 1+iz/z_0$, $z_0 = kw^2/2$ is the Rayleigh length, k is the wavenumber of light, w is the waist radius of the Gaussian beam, (r, φ, z) are the cylindrical coordinates, z is the propagation distance, α is the scaling factor that defines the width of the Bessel function $J_n(x)$, n is an integer number equal to the topological charge of the optical vortex. Using the following series of the Bessel functions (Sect. 5.7.8 in [25]):

$$\sum_{m=0}^{\infty} \frac{t^{2m+n}}{m!(m+n)!} J_{2m+n}(x) = J_n\left(x \sin\left(\frac{1}{2}\arcsin\frac{4t}{x}\right)\right) J_n\left(x \cos\left(\frac{1}{2}\arcsin\frac{4t}{x}\right)\right), \tag{4.19}$$

we derive the complex amplitude of a new family of Bessel-Bessel-Gaussian vortex laser beams (BBG beams):

$$\begin{aligned} BBG_n(r,\varphi,z) &= \frac{1}{q}\exp\left(-\frac{i\alpha^2 z}{2kq} - \frac{r^2}{w^2 q}\right) \sum_{m=0}^{\infty} \frac{(ce^{i\varphi})^{2m+n}}{m!(m+n)!} J_{2m+n}\left(\frac{\alpha r}{q}\right) \\ &= \frac{1}{q}\exp\left(-\frac{i\alpha^2 z}{2kq} - \frac{r^2}{w^2 q}\right) \\ &\times J_n\left(\frac{1}{q}\sqrt{\frac{\alpha r}{2}\left(\alpha r - \sqrt{\alpha^2 r^2 - 16c^2 q^2 e^{2i\varphi}}\right)}\right) \\ &\times J_n\left(\frac{1}{q}\sqrt{\frac{\alpha r}{2}\left(\alpha r + \sqrt{\alpha^2 r^2 - 16c^2 q^2 e^{2i\varphi}}\right)}\right), \end{aligned} \tag{4.20}$$

with a parameter c, which we suppose real and positive, while, of course, it can be chosen to be any complex number. The reason is that the replacement $c = |c| \to |c|e^{i\theta}$ leads to the beam (4.20) rotation over the angle $-\theta$ due to the presence of the term $c^2 e^{2i\varphi}$ in both Bessel factors. The complex amplitude of BBG beams (4.20) is written explicitly for an arbitrary propagation distance z. Tending z to infinity allows obtaining the complex amplitude of BBG beams in the far field. According to Eq. (4.20), BBG beams devoid of rotational symmetry, but they are axially symmetric since a replacement of the azimuthal angle φ by the angle $\varphi + \pi$ does not change the amplitude (4.20). If we suppose $\varphi = \pi p$ in the series (4.20), $p = 0,1,2\ldots$, then the series coefficients are all positive for different indices m, i.e. all BG beams in the series are summed up in phase, thus yielding a high intensity on the horizontal axis. On the vertical axis, i.e., when $\varphi = \pi/2 + \pi p$, the coefficients are of different signs and neighboring BG beams are summed in antiphase, so that the intensity on the vertical axis is lower than on the horizontal one. Tending the radial coordinate to zero ($r \to 0$) and employing approximations of the Bessel functions at small argument,

$$J_n(x << 1) \approx \frac{1}{n!}\left(\frac{x}{2}\right)^n, \tag{4.21}$$

we obtain the amplitude of the BBG beam near the intensity zero on the optical axis:

$$\begin{aligned}
BBG_n(r \ll 1, \varphi, z) &\approx \frac{1}{(n!)^2 q} \exp\left(-\frac{i\alpha^2 z}{2kq} - \frac{r^2}{w^2 q}\right) \\
&\times \left(\frac{\alpha r}{8q^2}\left(\alpha r - \sqrt{\alpha^2 r^2 - 16c^2 q^2 e^{2i\varphi}}\right)\right)^{n/2} \\
&\times \left(\frac{\alpha r}{8q^2}\left(\alpha r + \sqrt{\alpha^2 r^2 - 16c^2 q^2 e^{2i\varphi}}\right)\right)^{n/2} \\
&= \frac{1}{(n!)^2 q} \exp\left(-\frac{i\alpha^2 z}{2kq} - \frac{r^2}{w^2 q}\right)\left(\frac{\alpha r}{8q^2}\right)^n (4cqe^{i\varphi})^n \\
&= \frac{1}{(n!)^2 q} \exp\left(-\frac{i\alpha^2 z}{2kq} - \frac{r^2}{w^2 q}\right)\left(\frac{\alpha c}{2|q|} \cdot re^{i(\varphi-\psi)}\right)^n, \qquad (4.22)
\end{aligned}$$

where $\psi = \arctan(z/z_0)$.

Equation (4.22) reveals that the amplitude of the beam (4.20) near the intensity zero on the optical axis describes an optical vortex with the topological charge n. It is also seen from Eq. (4.22) that the rotation angle of the intensity pattern increases counterclockwise upon beam propagation, proportionally to the Gouy phase: $\varphi = \psi$. If the radial coordinate tends to zero ($r \to 0$), then only the term $m = 0$ in the series (4.20) is significant and Eq. (4.20) reduces to the following form

$$\mathrm{BBG}_n(r << 1, \varphi, z) \approx \frac{1}{qn!} \exp\left(-\frac{i\alpha^2 z}{2kq} - \frac{r^2}{w^2 q}\right)\left(ce^{i\varphi}\right)^n J_n\left(\frac{\alpha r}{q}\right). \qquad (4.23)$$

This expression is valid also for small values of the parameter c. From Eq. (4.23), near the central intensity zero, the BBG beam coincides with the BG beam (4.18), if we select $c = (n!)^{1/n}$. As seen from Eq. (4.20) and especially from Eq. (4.23), if $c = 0$ then the BBG beam amplitude reduces to zero. Equation (4.20) also indicates that in the initial plane, when $z = 0$ and $q(z) = 1$, the arguments of the Bessel functions have real values at certain azimuthal angles: $\varphi = \pi p/2$, $p = 0,1,2,\ldots$ Thus, the arguments of the both Bessel functions in Eq. (4.3) are real on the Cartesian axes x and y and at certain values of the radial coordinate r can be equal to the zeros of the Bessel functions $J_n(\gamma_{n,p}) = 0$, $p = 1, 2, 3, \ldots$ Making the arguments of the Bessel functions on the Cartesian axes in the initial plane equal to the real zeros of the Bessel functions, we get ($\varphi = \pi p/2$):

$$\alpha r\left(\alpha r \pm \sqrt{\alpha^2 r^2 \pm 16c^2}\right) = 2\gamma_{n,p}^2. \qquad (4.24)$$

The sign '−' in the radicand in Eq. (4.24) is for the axis x, whereas the sign '+' is for the axis y. The plus and minus signs before the square root in Eq. (4.24) correspond to different Bessel functions in Eq. (4.20). On the axis y, the sign '+' should be chosen in the radicand in Eq. (4.24). Thus, since the l.h.s. of Eq. (4.24) should be positive, the sign '+' before the square root should be selected. This allows unambiguous determining the coordinates of the Bessel function zeros on the vertical

Cartesian axis:

$$r_{y,p} = \pm \frac{\gamma_{n,p}^2}{\alpha \sqrt{4c^2 + \gamma_{n,p}^2}}. \tag{4.25}$$

The plus and minus signs in Eq. (4.25) denote the positive and negative coordinates of the intensity zeros of the BBG beam on the vertical axis in the initial plane. Since for the vertical axis, the plus sign should be chosen in the radicand in Eq. (4.24), the second term in Eq. (4.24) exceeds the first one, i.e., the outer radicand of the first Bessel function in Eq. (4.20) is negative on the vertical axis and the argument is thus purely imaginary:

$$\sqrt{\frac{\alpha r}{2}\left(\alpha r - \sqrt{\alpha^2 r^2 + 16c^2}\right)} = \pm i \sqrt{\frac{\alpha r}{2}\left(\sqrt{\alpha^2 r^2 + 16c^2} - \alpha r\right)}. \tag{4.26}$$

For the axis x, the minus sign should be chosen in the radicand in Eq. (4.24). Then, to make the radicand nonnegative, the following inequality should be satisfied:

$$r \geq \frac{4c}{\alpha}. \tag{4.27}$$

For large values of the asymmetry parameter c, $4c \gg \alpha$, this inequality indicates that the zeros of the Bessel function (intensity zeros) on the horizontal axis are located further from the origin than the zeros on the vertical axis. From Eq. (4.27) also follows that the intensity zeros on the horizontal axis reside further than the maximal intensity value, i.e., the radii of the two intensity maxima increase linearly with increasing parameter c and are close to $r_{\max} \approx \pm 4c/\alpha$. Coordinates of the intensity zeros on the horizontal axis are given by

$$r_{x,p} = \pm \frac{\gamma_{n,p}^2}{\alpha \sqrt{\gamma_{n,p}^2 - 4c^2}}. \tag{4.28}$$

The expression (4.28) is valid if the condition (4.27) is satisfied as well as the following condition of positive value of the radicand in Eq. (4.28):

$$\gamma_{n,p} > 2c. \tag{4.29}$$

We note that the intensity zeros (local optical vortices) on the Cartesian axes have alternating signs (topological charge is + 1 or − 1) so that the neighboring optical vortices compensate the influence of each other. This is the case with all the zeros excepting the degenerate zero at the origin, whose topological charge is n. The topological charge (TC) of the BBG beams increases with increasing asymmetry parameter $c > 0$. If the term $m = 0$ in the series (4.20) is the dominant, then the TC of the beam (4.20) is n. This is the case when the asymmetry parameter is small and approximately meets the condition $c < n$. When the parameter c increases, the TC of the series (4.20) is equal to the TC of the second term, i.e. $n + 2$. On further increasing the parameter c, the third term in the series (4.20) can have the largest magnitude (at small r) and then the TC of the superposition of all BG beams in the series becomes equal $n + 4$, and so on. If the parameter tends to infinity, the TC of the series (4.20) and of the BBG beam itself tends to infinity also. Upon propagation of the beam (4.20) in free space, the intensity zeros with the opposite signs on the Cartesian axes disappear (annihilate), and only the optical vortices near the origin remain.

4.2.2 Bessel-Bessel-Gaussian Beams Based Upon the Bessel-Gaussian Beams with Quadratic Argument

In 1999, Caron and Potvliege found a solution of the paraxial Helmholtz equation [20]:

$$\left(\frac{\partial^2}{\partial x^2} + \frac{\partial^2}{\partial y^2} + 2ik\frac{\partial}{\partial z}\right)E(x, y, z) = 0 \tag{4.30}$$

as a product of the Gaussian function by the Bessel function of a half-integer order and a squared argument (qBG beams) [20]:

$$\begin{aligned} qBG_n(r, \varphi, z) = {} & \frac{1}{\sqrt{q_+ q_-}} J_{n/2}\left(\frac{\beta r^2}{q_+ q_-}\right) \\ & \times \exp\left(\frac{i(1+\beta^2 w^4) z r^2}{z_0 w^2 q_+ q_-} - \frac{r^2}{w^2 q_+ q_-} + in\varphi\right), \end{aligned} \tag{4.31}$$

where $q_\pm = 1 \pm \beta w^2 z / z_0 + iz / z_0$ and β is a real parameter defining the scale of the Bessel function.

Based on the beams (4.31) and the series (4.19), qBBG beams can be constructed with a half-integer order:

$$\begin{aligned} \mathrm{qBBG}_n(r, \varphi, z) = {} & \frac{1}{\sqrt{q_+ q_-}} \exp\left(\frac{i(1+\beta^2 w^4) z r^2}{z_0 w^2 q_+ q_-} - \frac{r^2}{w^2 q_+ q_-}\right) \\ & \times \sum_{m=0}^{\infty} \frac{(ce^{2i\varphi})^{2m+n/2}}{m!\Gamma(m + n/2 + 1)} J_{2m+n/2}\left(\frac{\beta r^2}{q_+ q_-}\right) \\ = {} & \frac{1}{\sqrt{q_+ q_-}} \exp\left(\frac{i(1+\beta^2 w^4) z r^2}{z_0 w^2 q_+ q_-} - \frac{r^2}{w^2 q_+ q_-}\right) \\ & \times J_{\frac{n}{2}}\left(\frac{1}{q_+ q_-}\sqrt{\frac{\beta r^2}{2}\left(\beta r^2 - \sqrt{\beta^2 r^4 - 16c^2 (q_+ q_-)^2 e^{4i\varphi}}\right)}\right) \end{aligned}$$

$$\times J_{\frac{n}{2}}\left(\frac{1}{q_+q_-}\sqrt{\frac{\beta r^2}{2}\left(\beta r^2+\sqrt{\beta^2 r^4-16c^2(q_+q_-)^2 e^{4i\varphi}}\right)}\right). \tag{4.32}$$

The beam family (4.32), depending on a parameter c, which we suppose a real number, is of finite energy due to the Gaussian envelope. At $c = 0$, the beam (4.32) vanishes for $n \neq 0$, or reduces to the zero-order qBG beam (4.31) for $n = 0$. Similarly to Eq. (4.23), near the optical axis (at small r), the beam (4.32) reduces to the generating beam (4.31). Analysis of the series (4.32) reveals that at the angles $2\varphi = \pi p$, $p = 0, 1, 2\ldots$, the series coefficients are positive for all indices m, i.e. all qBG beams in the series (4.32) are summed in phase and the intensity should be large on the Cartesian axes. On the contrary, if $2\varphi = \pi/2 + \pi p$, the coefficients of the series (4.32) have alternating signs and the neighboring qBG beams are summed in antiphase and the intensity on the diagonals of the four quadrants is lower than on the Cartesian axes. Thus, it can be supposed that the qBBG beams have four local intensity maxima (four intensity lobes) on the Cartesian axes at a certain distance from the optical axis. Similarly to Eq. (4.24), we make the argument of the Bessel functions in Eq. (4.32) equal to the zeros of the Bessel function of a half-integer order in the initial plane ($z = 0$) on 8 rays $\varphi = \pi p/4$, $p = 0, 1, 2,\ldots$:

$$\beta r^2\left(\beta r^2 \pm \sqrt{\beta^2 r^4 \pm 16c^2}\right) = 2\gamma_{n/2,p}^2. \tag{4.33}$$

At angles $\varphi = \pi p/4$, $p = 1, 3, 5\ldots$, the plus sign should be chosen in the radicand and thus the Bessel function zeros are located on the four diagonals in points with the following coordinates:

$$r_{x,y,p} = \pm\frac{\gamma_{n/2,p}^2}{\sqrt{\beta}\cdot\sqrt[4]{\gamma_{n/2,p}^2+4c^2}}. \tag{4.34}$$

For odd values p, i.e. on the rays $\varphi = \pi p/4$, $p = 0, 2, 4\ldots$, there is the minus sign in the radicand of Eq. (4.33) and the square root has a real value if the following condition is satisfied, similar to Eq. (4.27):

$$r \geq 2\sqrt{\frac{c}{\beta}}. \tag{4.35}$$

In Eq. (4.35), the parameter c should be positive. Coordinates of Bessel functions zeros in Eq. (4.32) should be sought with accounting the condition (4.35) on the Cartesian axes at the following points:

$$r_{x,p} = \pm\frac{\gamma_{n/2,p}^2}{\sqrt{\beta}\cdot\sqrt[4]{\gamma_{n/2,p}^2-4c^2}}. \tag{4.36}$$

Since the expression (4.36) should be real, a following condition should be satisfied, similar to Eq. (4.29): $\gamma_{n/2,p} > 2c$. These zeros of the Bessel functions in Eq. (4.32), lying on the Cartesian axes and on the diagonals (on 8 rays), yield unitary-charged optical vortices with alternating sings $+1$ and -1. Therefore, these vortices do not affect the topological charge of the beam (4.32), equal to the topological charge of the central optical vortex n. Upon propagation in free space, the arguments of the Bessel functions in Eq. (4.32) become complex, and all noncentral optical vortices compensate each other and disappear. In the section of the beam (4.32) at $z > 0$, only the central optical vortex remains.

4.2.3 Bessel-Bessel-Gaussian Beams of the Second Type

The reference book [56] contains another series which is similar to the series (4.19):

$$\sum_{m=0}^{\infty} \frac{(n+1/2)_m t^{m+n}}{m!(2n+1)_m} J_{m+n}(2x) = 2^{2n} n! J_n\left(x+\sqrt{x^2-tx}\right) J_n\left(x-\sqrt{x^2-tx}\right), \tag{4.37}$$

with $(a)_k = \Gamma(a+k)/\Gamma(a)$ being the Pochhammer symbol. Based on the series (4.37) and the BG beam (4.18), BBG beams of the second type (BBG2 beams) can be constructed:

$$\begin{aligned} \mathrm{BBG2}_n(r,\varphi,z) &= \frac{1}{q}\exp\left(-\frac{i\alpha^2 z}{2kq}-\frac{r^2}{w^2 q}\right) \\ &\times \sum_{m=0}^{\infty} \frac{(n+1/2)_m (ce^{i\varphi})^{m+n}}{2^{2n} n! m! (2n+1)_m} J_{m+n}\left(\frac{\alpha r}{q}\right) \\ &= \frac{1}{q}\exp\left(-\frac{i\alpha^2 z}{2kq}-\frac{r^2}{w^2 q}\right) \\ &\times J_n\left(\frac{1}{2q}\left(\alpha r+\sqrt{\alpha^2 r^2-2\alpha cqre^{i\varphi}}\right)\right) \\ &\times J_n\left(\frac{1}{2q}\left(\alpha r-\sqrt{\alpha^2 r^2-2\alpha cqre^{i\varphi}}\right)\right). \end{aligned} \tag{4.38}$$

If $c = 0$, BBG2 beams vanish for $n \neq 0$, or reduce to the zero-order BG beam for $n = 0$. As seen from Eq. (4.38), the series yields real values in the initial plane only on the axis x. In this case, on the positive part of the axis ($\varphi = 2\pi p$, $p = 0, 1, 2\ldots$), all the terms of the series are summed 'in phase' and the intensity magnitude is the highest. On the negative part of the axis x ($\varphi = \pi p$, $p = 1, 3, 5\ldots$), the neighboring terms are summed in antiphase and the intensity is thus lower than that on the positive part of the axis x. Thus, intensity distribution in the initial plane has a single maximum at $x > 0$. To obtain the coordinates of the intensity zeros in the initial plane ($z = 0$), we make the arguments of the Bessel functions in Eq. (4.38) equal to the Bessel function zeros $\gamma_{n,p}$ on the negative part of the horizontal axis $x < 0$:

$$r_{x<0} = \frac{2\gamma_{n,p}^2}{\alpha(2\gamma_{n,p} + c)}. \tag{4.39}$$

On the positive part of the horizontal axis $x > 0$, the coordinates of the intensity zeros can be found only at the following condition, similar to Eq. (4.27):

$$r \geq \frac{2c}{\alpha}. \tag{4.40}$$

If the condition (4.40) is satisfied, coordinates of the intensity zeros are given by

$$r_{x>0} = \frac{2\gamma_{n,p}^2}{\alpha(2\gamma_{n,p} - c)}, \quad 2\gamma_{n,p} > c. \tag{4.41}$$

Comparison of Eqs. (4.39) and (4.41) indicates that if the asymmetry parameter c increases then the intensity zeros at $x < 0$ become closer to the origin, whereas the zeros at $x > 0$ move away from the origin. At the origin itself, on the optical axis, there is an intensity zero surrounded by an optical vortex with the topological charge n. The intensity zeros (4.39) and (4.41) on the horizontal axis are centers of optical vortices with the topological charges + 1 and − 1. When the beam propagates in space, these peripheral optical vortices "annihilate" since the argument of the Bessel functions in Eq. (4.38) becomes complex, whereas all the Bessel function zeros are real.

4.2.4 Numerical Simulation

Simulation of propagation of BBG beams (4.20)

Figure 4.4 illustrates the intensity and phase distributions of the BBG beam (4.20) for the following parameters: wavelength $\lambda = 532$ nm, waist radius $w_0 = 1$ mm, beam order $n = 4$, scaling factor of the BG beams $\alpha = k/1000$, asymmetry parameter $c = 5$. The intensity and phase distributions are shown for the following distances from the waist plane: $z = 0, 0.15z_0, 0.25z_0, 0.5z_0, z_0, 2z_0$. As seen in Fig. 4.4, after propagating over a distance which is much shorter than the Rayleigh length, the BBG beam is already rotated counterclockwise by almost 45°. At the Rayleigh length, the BBG beam (4.20) is rotated by almost 90°. On further propagation, the beam is slowly rotates up to 90°. Typically, vortex beams without radial symmetry of the intensity distribution rotate by 45° at a distance equal to the Rayleigh length [57, 58]. This is because the far field of the Bessel beam (4.20) appears at smaller distances that the Rayleigh length. The waist radius of the Gaussian beam can be treated as an aperture bounding the Bessel beam, and thus, after the distance $wk/\alpha = 1000w = 1$ m, the far field already appears for the Bessel beam, although the Rayleigh length is near $z_0 = 6$ m.

Shown in Fig. 4.5 are the intensity and phase distributions of the BBG beams (4.20) for the following parameters: wavelength $\lambda = 532$ nm, waist radius $w_0 = 1$ mm, topological charge $n = 4$, scaling factor $\alpha = k/1000$, asymmetry parameter $c = 1, 2, 3, 4, 5, 6, 7, 8$, propagation distance $z = z_0$. Half-size of the computation domain is $R = 10$ mm. As seen in Fig. 4.5, if the parameter c grows, then the BBG beam leads to increase of its topological charge. For instance, at c from 1 till 3, the TC is $n = 4$, at $c = 4$ and $c = 5$, the TC equals $n + 2 = 6$, at $c = 6$ and $c = 7$, the TC is $n + 4 = 8$, whereas at $c = 8$, the TC is equal to $n + 6 = 10$. Thus, we demonstrated that changing the parameter c allows controlling the topological charge of the BBG beams (4.20), changing it by an even number. Besides, Fig. 4.5 indicates that the distance between the two intensity maxima on the horizontal axis in the initial plane increases proportionally to the value of c, which is consistent with the theory (4.40). The distance from the origin to one of the crescents in the first column of Fig. 4.5 can be estimated by knowing the first zeros of the first Bessel functions present in Eq. (4.20). For example, at $c = 1$, the main contribution into the series (4.20) is given by the first term, proportional to the 4th-order Bessel function. The first zero of the 4th-order Bessel function is equal to 7.3. Then, making the argument αr of the Bessel function in Eq. (4.20) equal to the half of the first zero, we get an approximate distance to the crescent – 310 μm (the exact distance is 440 μm). For larger values c, the first zeros of the higher-order Bessel function should be chosen. The exact distances to the crescent in the initial plane in Fig. 4.5 (first column) for different values c are as follows: 479 μm ($c = 2$), 537 μm ($c = 3$), 655 μm ($c = 4$), 772 μm ($c = 5$). The radius of the inhomogeneous intensity ring in Fig. 4.5a3 is increased compared to the intensity radius in the waist (Fig. 4.5a1) nearly 10 times. The radius of this ring can be estimated by the formula $z_0\alpha/k = z_0/1000 = 6.2$ mm. This is consistent with Fig. 4.5a3 (ring radius of 5.8 mm).

Simulation of propagation of qBBG beams (4.32)

For simulation of the qBBG beams (4.32), we used the following parameters: waist radius of the Gaussian beam $w_0 = 1$ mm, topological charge $n = 4$, wavelength $\lambda = 532$ nm, scaling factor of the Bessel function $\beta = 10^{-6}k_0^2$ ($k_0 = 2\pi/\lambda$ is the wavenumber), and the values of the asymmetry parameter are: $c = 1, 1.5, 2$. Since the beams (4.32) are highly diverging, the intensity was computed at a small distance from the initial plane, ten times shorter than the Rayleigh length: $z = z_0/10$. The half-size of the images is $R = 5$ mm. Half-size in the initial plane is equal to $R = 1$ mm. Shown in Figs. 4.6, 4.7 and 4.8 are beam intensity distributions in the initial plane (the initial beam has a radius nearly equal to 0.25 mm) (Figs. 4.6a, 4.7a, 4.8a), and two intensity distributions at a distance of $z = z_0/10$, computed by two different ways: by using the integral Fresnel transform (Figs. 4.6b, 4.7b, 4.8b), and by the series (4.32) of the Bessel functions with a quadratic argument (Figs. 4.6c, 4.7c,

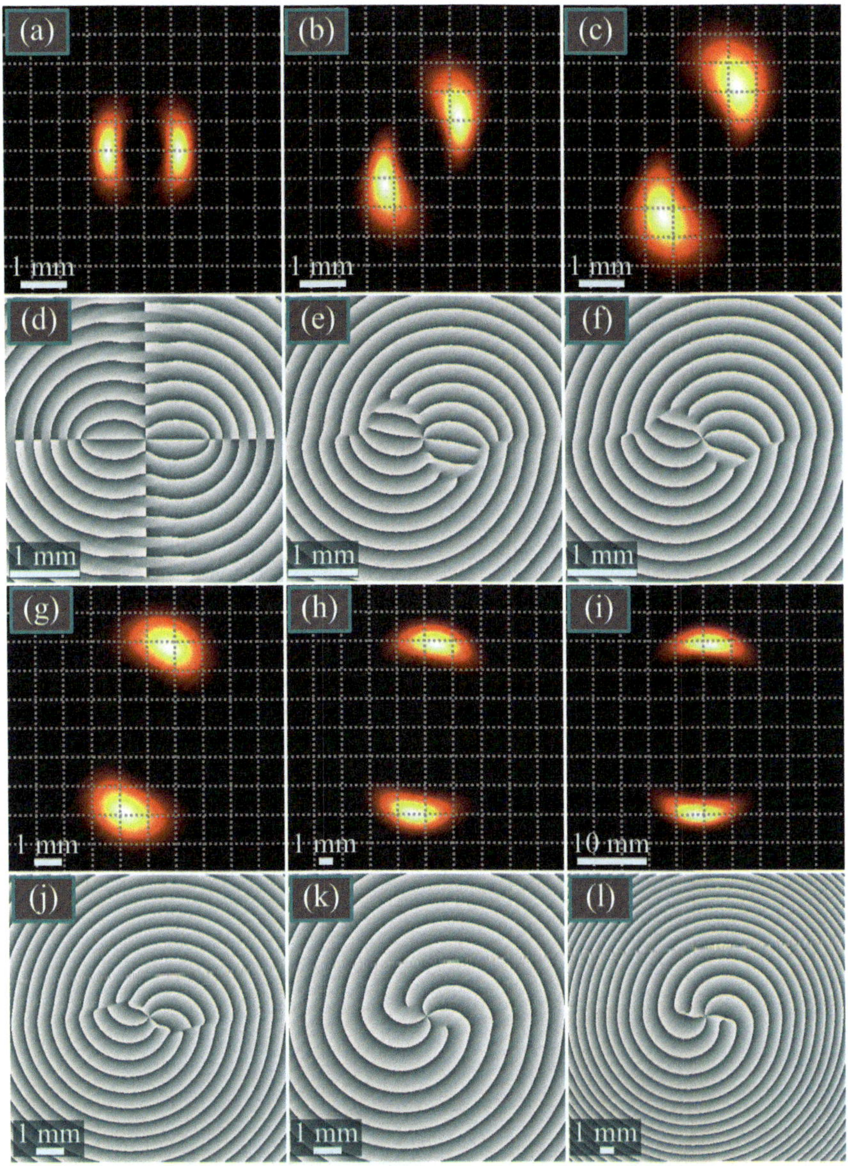

Fig. 4.4 Intensity and phase distributions of BBG beams, computed by Eq. (4.20), at different distances from the beam waist: $z = 0$, $0.15z_0$, $0.25z_0$, $0.5z_0$, z_0, $2z_0$

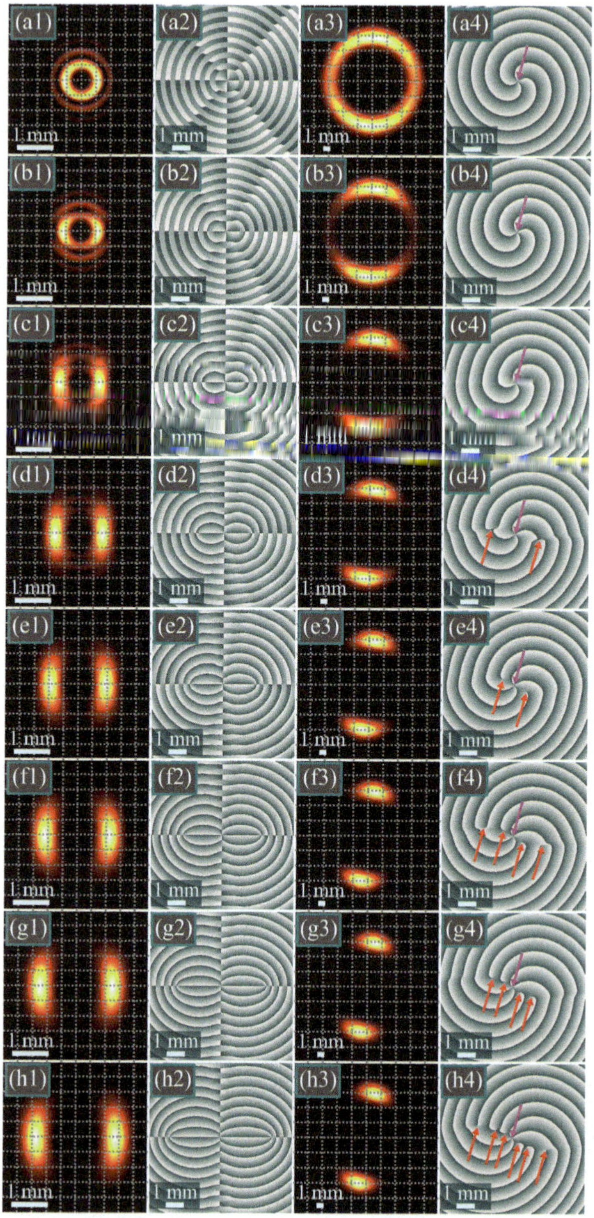

Fig. 4.5 Intensity (columns 1 and 3) and phase (columns 2 and 4) distributions of the BBG beam (4.20) in the initial plane (columns 1 and 2) and at the Rayleigh distance (columns 2 and 4) for different values of the asymmetry parameter c: 1(row 1), 2(row 2), 3 (row 3), 4 (row 4), 5 (row 5), 6 (row 6), 7 (row 7) and 8 (row 8). Pink arrows (column 4) show the central nth-order vortex, while the red arrows show off-axis vortices with the TC of + 1

4.8c). Figures 4.6, 4.7 and 4.8 were obtained for different values of the asymmetry parameter c: 1 (Fig. 4.6), 1.5 (Fig. 4.7), 2 (Fig. 4.8). According to Figs. 4.6, 4.7 and 4.8, even at a distance ten times shorter than the Rayleigh length, the beam size becomes nearly 10 times wider. Thus, despite the beam (4.36) is paraxial since it has the Gaussian envelope, it diverges much stronger then the enveloping Gaussian beam.

The distances between the intensity maxima (along the horizon) in the initial plane are equal to 303 μm (Fig. 4.6a), 322 μm (Fig. 4.7a), 342 μm (Fig. 4.8a). At the distance $z_0/10$, the distances between the maxima are 4221 μm (Fig. 4.6b, c), 4396 μm (Fig. 4.7b, c), 4669 μm (Fig. 4.8b, c). Thus, along such a small propagation distance, the beam width increases 13.9 ($c = 1$), 13.6 ($c = 1{,}5$), 13.7 ($c = 2$) times.

Figure 4.9 illustrates phase distributions in the initial plane for the beams from Figs. 4.6, 4.7 and 4.8. Shown in Fig. 4.6a is the phase of the beam with $c = 1$ (Fig. 4.3a), Fig. 4.6b shows the phase of the beam with $c = 1.5$ (Fig. 4.7a), and Fig. 4.9c depicts the phase of the beam with $c = 2$ (Fig. 4.8a). According to Fig. 4.9, as the theory predicts, the zeros (intensity zeros) on the Cartesian axes move away from the origin with increasing values c (4.36), whereas the zeros on the diagonals move closer to the origin (4.34). The first ring in Fig. 4.9a has a shape of a square

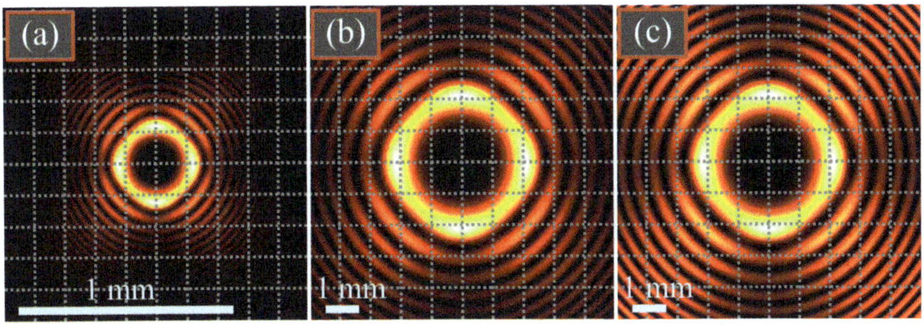

Fig. 4.6 Intensity in the initial plane, computed as a product of the Bessel functions (**a**), as well as the intensities at a distance $z = z_0/10$ (**b**, **c**), obtained by using the Fresnel transform (**b**) and by the series (4.32) of qBG beams (**c**), at $c = 1$

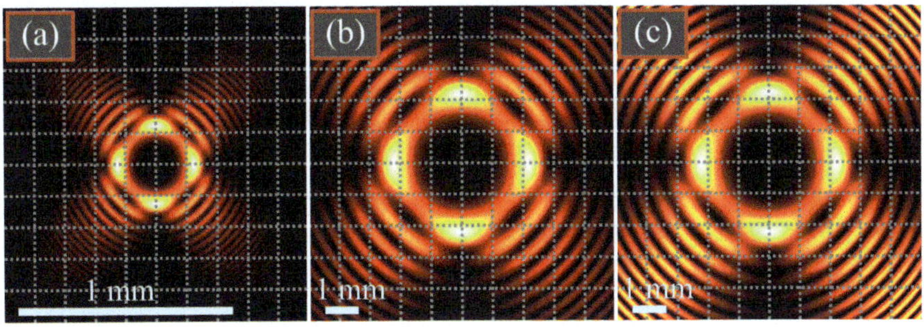

Fig. 4.7 Same as Fig. 4.6 but for $c = 1.5$

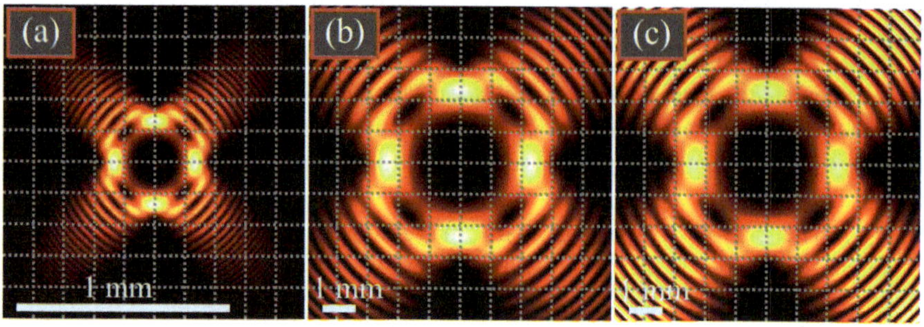

Fig. 4.8 Same as Fig. 4.6 but for $c = 2$

with the sides convex outside, but the same first ring in Fig. 4.7, has a square shape with the sides concave inside (pillow). In the beam center (on the optical axis), there is an optical vortex with the topological charge $n = 4$.

As seen in Figs. 4.6, 4.7 and 4.8, the BBG2 beams are not rotated upon propagation and their transverse intensity distribution changes weakly. Nevertheless, these beams cannot be considered as propagation-invariant (structurally stable) since the far-field intensity distributions of these beams have the shape of concentric light rings. This is seen from the series in Eq. (4.32), since the divergence of the Bessel beams in the series grows with increasing of their topological charge, whereas the amplitudes of these beams (weight coefficients in the series (4.32)) decrease. Thus, the brightest light ring in the far field is formed by the Bessel beam with the topological charge of n, which is the term $m = 0$ in the series (4.32). Figure 4.10 illustrates the intensity distribution of the BBG2 beam (4.32) with $c = 1$ at a distance equal to half Rayleigh length ($z = z_0/2$). It is seen that even at this distance, the beam with a shape of round-corners square (Fig. 4.6) turns into a ring-shape beam surrounded by side rings.

Now we estimate the divergence of the beam (4.32). The qBG beam with the topological charge n, which is the term $m = 0$ in the series (4.32), has the initial radius equal to that of the conventional BG beams, i.e. $r_0 \simeq w_0\sqrt{n/2}$. The qBG beam

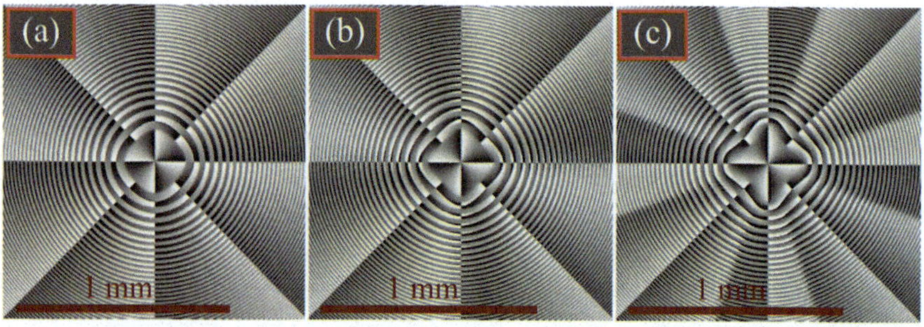

Fig. 4.9 Phase distributions in the initial plane of the beams shown in Figs. 4.6a, 4.7 and 4.8a with the parameter c equal to 1 (**a**), 1.5 (**b**), 2 (**c**)

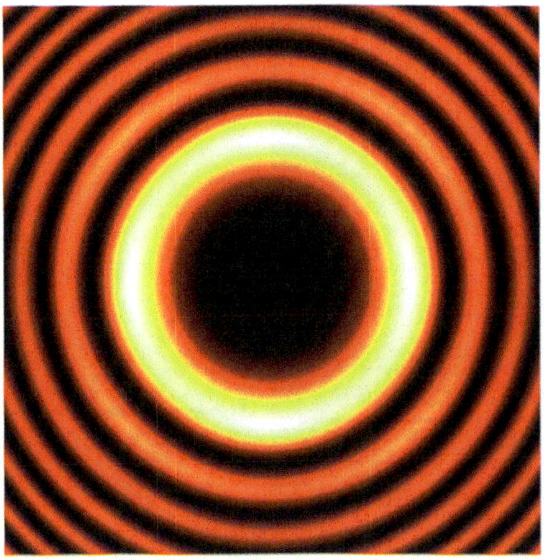

Fig. 4.10 Intensity distribution of the beam (4.32) at a distance $z = z_0/2$ from the waist plane at $c = 1$. The frame size is 10×10 mm

with an argument βr^2 propagates as after a spherical lens with the focal distance $f = k/(2\beta)$, i.e. it diverges after passing the focus plane and, at a distance $z = z_0/10$, its radius r_z increases $(z_0/10 - f)/f$ times compared to the initial radius r_0. Thus, we can estimate the divergence of the beam (4.32) as a ratio of its radii:

$$\frac{r_z}{r_0} = \frac{z_0}{10f} - 1 = \frac{w^2\beta}{10} - 1 = \frac{(kw)^2}{10^7} - 1 \approx 14.$$

This value is slightly higher than the value obtained in numerical simulation, 13.6–13.9.

Simulation of propagation of BBG2 beams (4.38)

For simulating the beam (4.38) we used the following parameters: wavelength $\lambda = 532$ nm, waist radius $w_0 = 1$ mm, topological charge $n = 4$, scaling factor of the constituent BG beams $\alpha = k/200$, asymmetry parameter $c = 1$, propagation distances $z = 0, 0.1z_0, 0.2z_0$, half-size of the computation area $R = 10$ mm. Intensity distributions at several distances are shown in Fig. 4.11, where the initial distribution is zoomed 10 times.

As seen from comparison of the intensities in Fig. 4.11, the beam is increased by size upon propagation, rotated by 90° and "looses" the side lobes. It is known for the conventional Bessel beam that its far-field intensity distribution has the shape of a ring without the side lobes. The far-field BG beam has also the shape of a ring, but it is thicker. Thus, if two coaxial Bessel beams with different orders are superimposed, then in the far field, all they generate a single ring with the intensity distribution along this ring being dependent on the weight coefficients in the sum

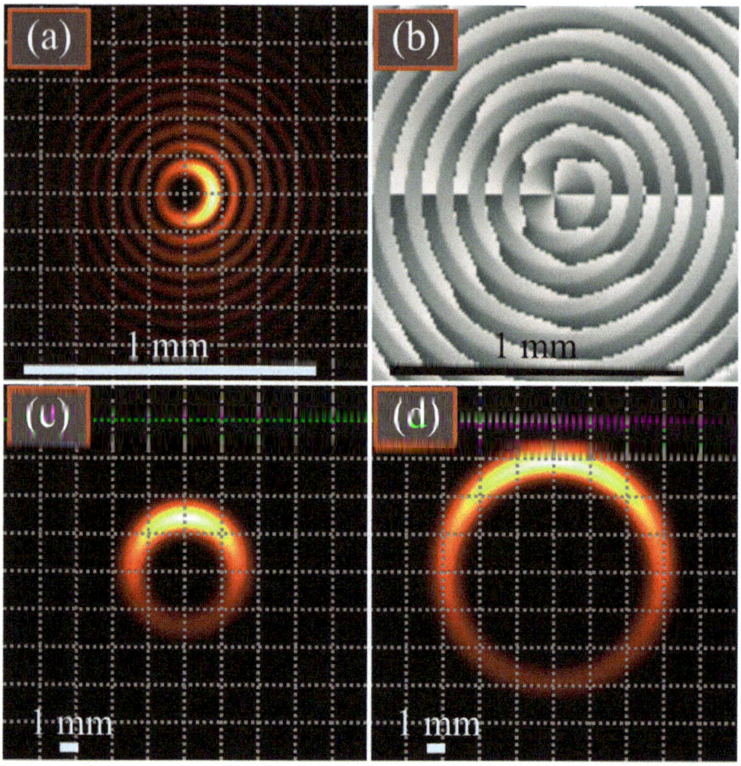

Fig. 4.11 Intensity distributions of the BBG2 beam (4.38) at some propagation distances z: 0 (**a**), $0.1z_0$ (**c**), $0.2z_0$ (**d**), as well as the phase distribution in the initial plane (**b**)

(4.38). If such a superposition has in the initial plane an intensity maximum with the center on the horizontal axis as in Fig. 4.11a, then, in the far field, the center of the intensity maximum should be on the vertical axis as in Fig. 4.11c, d. Thus, the intensity distribution should be rotated by 90°. This follows from the argument of each Bessel function in the series (4.38), which acquires in the far field an imaginary unit with minus, since the argument of the complex-valued Gouy phase $q(z)$ equals $\pi/2$. Therefore, for each Bessel function, we get in the far field: $J_n(-ix) = (-i)^n I_n(x)$ with $I_n(x)$ being the modified Bessel function. Then, in the sum (4.38), we get the coefficients $\left(ce^{i(\varphi-\pi/2)}\right)^{m+n}$ instead of $\left(ce^{i\varphi}\right)^{m+n}$. Thus, the form of the intensity distribution along the light ring in the far field does not change if the initial pattern is rotated 90° counterclockwise. Now we estimate the divergence of the beam (4.38). We note that the far field for the Bessel-Gaussian beam is formed in a different way than for the Gaussian beam. It is formed earlier, at $z > z_1 = wk/\alpha$, rather than at $z > z_0$. Since $\alpha = k/200$, then $z_1 = 200w = 200$ mm, whereas the Rayleigh length for the beam from Fig. 4.11 is equal to $z_0 = kw^2/2 \sim 6000$ mm. Thus, even the distance $0.1z_0 = 600$ mm (Fig. 4.11c, d) is a far field for the beams (4.38). The ring (crescent) radius R_1 in Fig. 4.11c can be estimated from an expression $R_1 \sim z \tan\theta$, where θ is the tilt angle of the conical rays that form the Bessel beams in the series (4.38). Since $\tan\theta \sim \sin\theta = \alpha/k$, then $R_1 = z\alpha/k = (z_0/10)/200 = 3$ mm. An exact computation shows that the distance from the optical axis to the crescent center in Fig. 4.11c is 2.874 mm.

Shown in Fig. 4.12 is the intensity of the BBG2 beam (4.38) at a distance $z = 0.04z_0$. An unusual feature of this intensity distribution is that it contains two beam types with different scales. At the origin, there is a small intensity distribution with a shape of a crescent, whose center is on the horizontal axis. Besides, an inhomogeneous much larger light ring appears, that also has a crescent shape, but with a center on the vertical axis. Presence of two different-scale beams in a single optical field in Fig. 4.12 can be explained by the nature of BG beams. The Gaussian beam acts as a soft aperture that bounds the aperture of the Bessel beam. Therefore, the Bessel beam propagates almost diffraction-free, saving its initial size, up to a distance nearly equal to $z_1 = wk/\alpha$. This diffraction-free beam is generated by a conical wavefront and, at a distance $z = 0.04z_0 = 240$ mm, those rays come to the optical axis that start from the "edge of the initial aperture", i.e., from a distance to the axis equal to the waist radius w. The rays that cross the optical axis at a shorter distance z, diverge and generate at a distance $z = 0.04z_0$ another beam type, i.e., the far-field beam (a crescent without sidelobes, rotated by 90° as in Fig. 4.12). On further propagation, diffraction-free part of the beam disappears and only the beam part in the far field remains, as in Fig. 4.11c, d. The size of the diffraction-free part of the beam (4.38), equal to the beam size in the initial plane (Fig. 4.11a), can be estimated by a half distance to the first zero of the 4th-order Bessel function. The zero of the 4th-order Bessel function is near 7.3, and thus the distance to the crescent in Fig. 4.11a is $r = 3.6/\alpha = 3.6 \times 200/k = 61$ μm. The exact distance is 88 μm.

The investigated BBG beams can be generated by a spatial light modulator with adopting well-known methods of encoding an amplitude-and-phase function (4.20)

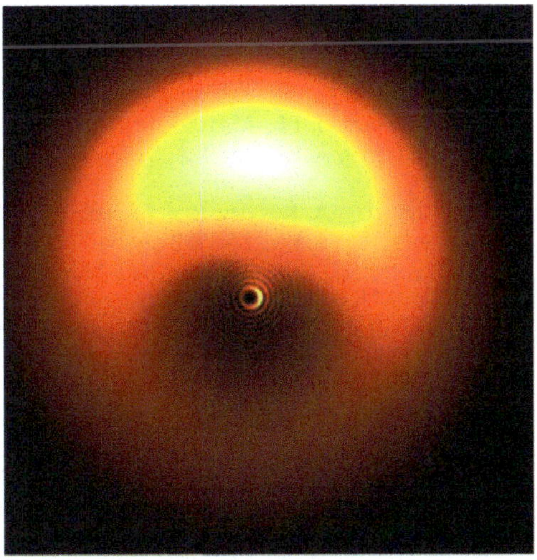

Fig. 4.12 Intensity distribution of the BBG beam (4.38) at a distance $z = 0.04z_0$. The frame size is 6 × 6 mm

into a phase-only function. Anomalously fast transverse rotation of the two-petal intensity pattern can be applied for improving the longitudinal resolution of an optical microscope, since a shift of a light source along the optical axis leads to the transverse rotation of the two intensity petals [59]. The faster is rotation of the two intensity petals, the higher is the longitudinal resolution of the optical microscope. A two-petal beam with a shape of two 'crescents' can be used for optical trapping biological cells by their edges and then moving them without damaging by the high-power irradiance of the Gaussian beam center. Smooth change of the beam parameter c allows stretching cells, since with increasing parameter c, not only the beam topological charge increases, but also the distance between the two intensity petals.

References

1. F. Gori, G. Guattari, C. Padovani, Bessel-gauss beams. Opt. Commun. **64**, 491–495 (1987)
2. I.P. Lukin, Coherence of Bessel-Gaussian beams propagating in a turbulent atmosphere. Atmos. Oceanic Opt. **31**, 49–59 (2018)
3. C. Bao-Suan, P. Ji-Xiong, Propagation of Gauss-Bessel beams in turbulent atmosphere. Chin. Phys. B **18**, 1033 (2009)
4. X. Yue, X. Ge, Y. Lyu, R. Zhao, B. Wang, K. Han, W. Zhang, Z. Man, S. Fu, Mean intensity of lowest order Bessel-Gaussian beams with phase singularities in turbulent atmosphere. Optik **219**, 165215 (2020)
5. H.T. Eyyuboğlu, F. Hardalaç, Propagation of modified Bessel-Gaussian beams in turbulence. Opt. Laser Technol. **40**, 343–351 (2008)
6. W. Nelson, J.P. Palastro, C.C. Davis, P. Sprangle, Propagation of Bessel and airy beams through atmospheric turbulence. J. Opt. Soc. Am. A **31**, 603–609 (2014)
7. P. Birch, I. Ituen, R. Young, C. Chatwin, Long-distance Bessel beam propagation through Kolmogorov turbulence. J. Opt. Soc. Am. A **32**, 2066–2073 (2015)
8. L. Zhu, X. Xin, H. Chang, X. Wang, Q. Tian, Q. Zhang, R. Gao, B. Liu, Security enhancement for adaptive optics aided longitudinal orbital angular momentum multiplexed underwater wireless communications. Opt. Express **30**, 9745–9772 (2022)
9. M.S. Kulya, V.A. Semenova, V.G. Bespalov, N.V. Petrov, On terahertz pulsed broadband Gauss-Bessel beam free-space propagation. Sci. Rep. **8**, 1390 (2018)
10. L. Yu, Y. Zhang, J. Wang, Propagation of asymmetric Bessel mode in turbulent atmosphere. IEEE Photonics J. **14**, 7354106 (2018)
11. V.V. Kotlyar, A.A. Kovalev, V.A. Soifer, Asymmetric Bessel modes. Opt. Lett. **39**, 2395–2398 (2014)
12. V.V. Kotlyar, A.A. Kovalev, R.V. Skidanov, V.A. Soifer, Asymmetric Bessel-Gauss beams. J. Opt. Soc. Am. A **31**, 1977–1983 (2014)
13. M. Mazilu, S.J. Stevenson, F. Gunn-Moore, K. Dholakia, Light beats the spread: "non-diffracting" beams. Laser Photonics Rev. **4**, 529–547 (2010)
14. S.N. Khonina, V.V. Kotlyar, R.V. Skidanov, V.A. Soifer, K. Jefimovs, J. Simonen, J. Turunen, Rotation of microparticles with Bessel beams generated by diffractive elements. J. Mod. Opt. **51**, 2167–2184 (2004)
15. R. Jáuregui, S. Hacyan, Quantum-mechanical properties of Bessel beams. Phys. Rev. A **71**, 033411 (2005)
16. M. McLaren, T. Mhlanga, M.J. Padgett, F.S. Roux, A. Forbes, Self-healing of quantum entanglement after an obstruction. Nat. Commun. **5**, 3248 (2014)
17. X. Chu, Q. Sun, J. Wang, P. Lü, W. Xie, X. Xu, Generating a Bessel-Gaussian beam for the application in optical engineering. Sci. Rep. **5**, 18665 (2016)

18. T. Yu, H. Xia, W. Xie, G. Xiao, H. Li, The generation and verification of Bessel-Gaussian beam based on coherent beam combining. Results Phys. **16**, 102872 (2019)
19. R. Vasilyeu, A. Dudley, N. Khilo, A. Forbes, A, Generating superpositions of higher-order Bessel beams. Opt. Express **17**, 23389–23395 (2009)
20. C.F.R. Caron, R.M. Potvliege, Bessel-modulated Gaussian beams with quadratic radial dependence. Opt. Commun. **164**, 83–93 (1999)
21. A. Belafhal, L. Dalil-Essakali, Collins formula and propagation of Bessel-modulated Gaussian light beams through an ABCD optical system. Opt. Commun. **177**, 181–188 (2000)
22. M.A. Bandres, D. Lopez-Mago, J.C. Gutiérrez-Vega, Higher-order moments and overlaps of rotationally symmetric beams. J. Opt. **12**, 015706 (2010)
23. V.V. Kotlyar, A.A. Kovalev, E.S. Kozlova, A.A. Savelyeva, S.S. Stafeev, New type of vortex laser beams: squared Laguerre-Gaussian beam. Optik **270**, 169916 (2022)
24. V.V. Kotlyar, E.G. Abramochkin, A.A. Kovalev, A.A. Savelyeva, Product of two Laguerre-Gaussian beams. Photonics **9**, 496 (2022)
25. A.P. Prudnikov, Y.A. Brychkov, O.I. Marichev, *Integrals and Series: Special Functions* (Gordon and Breach Science Publishers, New York, 1986)
26. E.G. Abramochkin, V.V. Kotlyar, A.A. Kovalev, Double and square Bessel-Gaussian beams. Micromachines **14**, 1029 (2023)
27. S. Zhang, M. Elsayed, R. Peng et al., Reconfigurable multi-component micromachines driven by optoelectronic tweezers. Nat. Commun. **12**, 5349 (2021)
28. M. Li, X. Chen, S. Yan, Y. Zhang, B. Yao, Enatioselective rotation of chiral particles by azimuthally polarized beams. Adv. Photonics Res. **3**, 2200117 (2022)
29. M. Mazilu, K. Dholakia, Trapping and rotation of particles in light fields with embedded optical vortices, in *In Twisted Photons*, eds. by J.P. Torres, L. Torner (2011)
30. J. Durnin, Exact solutions for nondiffracting beams. I. The scalar theory. J. Opt. Soc. Am. A **4**, 651–654 (1987)
31. S.N. Khonina, N.L. Kazanskiy, S.V. Karpeev, M.A. Butt, Bessel beam: Significance and applications—a progressive review. Micromachines **11**, 997 (2020)
32. R. Stoian, M.K. Bhuyan, G. Zhang, G. Cheng, R. Meyer, F. Courvoisier, Ultrafast Bessel beams: advanced tools for laser materials processing Adv. Opt. Tech. **7**, 165–174 (2018)
33. M. Duocastella, C.B. Arnold, Bessel and annular beams for materials processing. Laser Photonics Rev. **6**, 607–621 (2012)
34. P.K. Velpula, M.K. Bhuyan, F. Courvoisier, H. Zhang, J.P. Colombier, R. Stoian, Spatio-temporal dynamics in nondiffractive Bessel ultrafast laser nanoscale volume structuring. Laser Photonics Rev. **10**, 230–244 (2016)
35. J. Arlt, V. Garcés-Chávez, W. Sibbett, K. Dholakia, Optical micromanipulation using a Bessel light beam. Opt. Commun. **197**, 239–245 (2001)
36. G. Milne, K. Dholakia, D. McGloin, K. Volke-Sepulveda, P. Zemánek, Transverse particle dynamics in a Bessel beam. Opt. Express **15**, 13972–13987 (2007)
37. R.A. Suarez, L.A. Ambrosio, A.A. Neves, M. Zamboni-Rached, M.R. Gesualdi, Experimental optical trapping with frozen waves. Opt. Lett. **45**, 2514–2517 (2020)
38. D. Rivero, V.S. de Angelis, C. Beli, M. Moreno, L.A. Ambrosio, P.W. Courteille, Hollow Bessel beams for guiding atoms between vacuum chambers: a proposal and efficiency study. J. Opt. Soc. Am. B **37**, 2660–2667 (2020)
39. J. Arlt, K. Dholakia, J. Soneson, E.M. Wright, Optical dipole traps and atomic waveguides based on Bessel light beams. Phys. Rev. A **63**, 063602 (2001)
40. Y. Liang, S. Yan, B. Yao, M. Lei, Direct observation and characterization of optical guiding of microparticles by tightly focused non-diffracting beams. Opt. Express **27**, 37975–37985 (2019)
41. F.O. Fahrbach, A. Rohrbach, Propagation stability of self-reconstructing Bessel beams enables contrast-enhanced imaging in thick media. Nature Commun. **3**, 632 (2012)
42. P.T. So, C.Y. Dong, B.R. Masters, K.M. Berland, Two-photon excitation fluorescence microscopy. Ann. Rev. Biomed. Eng. **2**, 399–429 (2000)
43. G. Thériault, Y. De Koninck, N. McCarthy, Extended depth of field microscopy for rapid volumetric two-photon imaging. Opt. Express **21**, 10095–10104 (2013)

44. N. Mphuthi, R. Botha, A. Forbes, Are Bessel beams resilient to aberrations and turbulence? J. Opt. Soc. Am. A **35**, 1021–1027 (2018)
45. S. Chen, S. Li, Y. Zhao, J. Liu, L. Zhu, A. Wang, J. Du, L. Shen, J. Wang, Demonstration of 20-Gbit/s high-speed Bessel beam encoding/decoding link with adaptive turbulence compensation. Opt. Lett. **41**, 4680–4683 (2016)
46. S. Li, J. Wang, Adaptive free-space optical communications through turbulence using selfhealing Bessel beams. Sci. Rep. **7**, 4323 (2017)
47. V.V. Kotlyar, A.A. Kovalev, D.S. Kalinkina, Fractional-order-Bessel Fourier-invariant optical vortices. Opt. Commun. **492**, 126974 (2021)
48. C.J.R. Sheppard, M.A. Porras, Comparison between the propagation properties of Bessel-Gauss and generalized Laguerre-Gauss beams. Photonics **10**, 1011 (2023)
49. V. Bagini, F. Frezza, M. Santarsiero, G. Schettini, G. Shirripa Spagnolo, Generalized Bessel-Gauss beams. J. Mod. Opt. **43**, 1155–1166 (1996)
50. H.F. Wang, F. Gan, High focal depth with a pure-phase apodizer. App. Opt. **40**, 5658–5662 (2001)
51. A. Mondal, A. Yevick, L.C. Blackburn, N. Kanellakopoulos, D.G. Grier, Projecting non-diffracting waves with intermediateplane holography. Opt. Express **26**, 3926–3931 (2018)
52. J.H. McLeod, The axicon: a new type of optical element. J. Opt. Soc. Am. **44**, 592–597 (2004)
53. R. Arimoto, C. Saloma, T. Tanaka, S. Kawata, Imaging properties of axicon in a scanning optical system. App. Opt. **31**, 6653–6657 (1992)
54. T. Breen, N. Basque-Giroux, U. Fuchs, I. Golub, Tuning the resolution and depth of field of a lens using an adjustable ring beam illumination. App. Opt. **59**, 4744–4749 (2020)
55. C.J.R. Sheppard, T. Wilson, Gaussian-beam theory of lenses with annular aperture. IEE J. Microwaves Opt. Acoust. **2**, 105–112 (1978)
56. Y.A. Brychkov, *Handbook of Special Functions: Derivatives, Integrals, Series and Other Formulas* (Chapman and Hall/CRC, 2008)
57. A.A. Kovalev, V.V. Kotlyar, Optical vortex beams with the infinite topological charge. J. Opt. **23**, 055601 (2021)
58. A.A. Kovalev, V.V. Kotlyar, Orbital angular momentum of generalized cosine Gaussian beams with an infinite number of screw dislocations. Optik **242**, 166863 (2021)
59. V.V. Kotlyar, E.G. Abramochkin, A.A. Kovalev, Bessel-Bessel-Gaussian vortex laser beams. J. Opt. **26**, 105601 (2024)

Chapter 5
Superposition of Helical Laser Beams

5.1 Topological Charge of a Superposition of Identical Parallel Laguerre-Gaussian Beams

Optical vortices constitute a great family of light fields, which is actively studied over 30 years [1]. The studies include various aspects, including direct generation in lasers [2], interaction with matter [3], propagation and focusing [4]. Light fields with optical vortices are usually characterized by the orbital angular momentum (OAM) [5] and the topological charge (TC) [6]. In a number of studies, the topological charge (TC) of a superposition of parallel optical vortices (OVs), and in particular, parallel Laguerre-Gaussian (LG) beams, was studied. This problem has been of interest since 2000, when the number and location of OVs in a superposition of two parallel Gaussian beams with embedded OVs were studied in [7]. In [7], a transcendental equation was obtained analytically for determining the position of OVs. However, it is applicable only for the case when the vortices in both beams are of the first order. It is also shown that when two beams are separated by a certain critical distance, negative-order vortices appear along with positive-order vortices. Later, in [8], using the analysis of forks in the interference pattern of two parallel LG beams, it was shown that when varying of the distance between the beams changes the arrangement of screw dislocations in the superposition. In addition, the same authors [9] showed that the number of vortices in the superposition of two parallel LG beams can change during propagation in space, although the total TC remains unchanged. In [10], the superposition of two off-axis optical vortices, but with orthogonal polarization, is studied. Instead of the dynamics of phase singularities, this paper studied the dynamics of polarization singularities and the position of C-points as a function of the distance between vortices, their TC, and the phase delay between them. In [11], the interference of two off-axis Gaussian beams with different curvature of the wave front is also considered. The conditions for vortex dipoles (two OVs of opposite orders) formation are obtained. In [12], the coherent and the incoherent superposition of two parallel partially coherent OVs is studied. It is shown that the

V. V. Kotlyar et al., *Helical Laser Beams*, https://doi.org/10.1007/978-3-031-92217-6_5

type of superposition, the distance between the beams, the propagation distance, and the coherence parameter affect the number and location of coherence vortices. The number of vortices, however, is determined only numerically. Study [13] considered OVs which are formed in a superposition of off-axis vortices while a nonlinear process of three-wave mixing. The number of vortices and their TCs have been established in some particular cases. In a recent article [14], the interaction of parallel Bessel-Gaussian beams is considered. The dependence of formation, annihilation and splitting of OVs on the displacement of the beams from the optical axis, on their TC, and on the phase difference between them is studied. It is shown that the total TC of such a composite field is not necessarily equal to the sum of the TCs of the composite beams. In [15], it was shown how to calculate the TC of a superposition of only two parallel LG beams. In particular, in [15] it was analytically shown that if two beams have the same TC, for example, m, then the superposition of such beams with arbitrary distance between them will also have a TC equal to m.

In this section, we generalize the results on a superposition of a finite number of parallel one-ring identical LG beams. And we will show that if the weight coefficients of such a superposition are real (that is, all LG beams have the same phase, but can have different amplitudes), then the superposition TC is equal to the TC of each beam, that is, m. It has already been proven earlier that the power-normalized OAM of such a superposition is also equal to the OAM of one LG beam in the superposition, that is, also m [16].

5.1.1 TC of the Superposition of Identical Parallel LG Beams in the Initial Plane

We consider a superposition of N parallel identical one-ring LG beams in the initial plane:

$$E_m(x,y)=\sum_{n=1}^{N}c_n\left(re^{i\varphi}-r_ne^{i\varphi_n}\right)^m\exp\left(-r^2-r_n^2+2rr_n\cos(\varphi-\varphi_n)\right),\qquad(5.1)$$

where (x, y) and (r, φ) are respectively the Cartesian and the polar coordinates ($x = r\cos\varphi$, $y = r\sin\varphi$).

The TC of each beam in (5.1) is equal to m, the waist radius is included in the radial variable r: r/w. We assume that the weights coefficient c_n in (5.1) are real numbers. Polar coordinates of beam centers are $(r_n,\ \varphi_n)$. . The TC of superposition (5.1) will be calculated using the Berry formula [6]:

$$TC=\frac{1}{2\pi}\lim_{r\to\infty}\operatorname{Im}\int_0^{2\pi}d\varphi\frac{\partial E(r,\varphi)/\partial\varphi}{E(r,\varphi)},\qquad(5.2)$$

where lim means the limit at $r \to \infty$, and Im is the imaginary part of a complex number.

Substituting (5.1) into (5.2), we obtain:

$$
\begin{aligned}
TC &= \frac{1}{2\pi} \lim_{r\to\infty} \operatorname{Im} \int_0^{2\pi} \left[\sum_{n=1}^{N} c_n \left(re^{i\varphi} - r_n e^{i\varphi_n}\right)^m e^{-r_n^2 + 2rr_n \cos(\varphi-\varphi_n)} \right. \\
&\quad \times \left. \left(\frac{imre^{i\varphi}}{re^{i\varphi} - r_n e^{i\varphi_n}} - 2rr_n \sin(\varphi - \varphi_n) \right) \right] \\
&\quad \left[\sum_{m=1}^{N} c_n \left(re^{i\varphi} - r_n e^{i\varphi_n}\right)^m e^{-r_n^2 + 2rr_n \cos(\varphi-\varphi_n)} \right]^{-1} d\varphi \\
&= m - \frac{2r}{2\pi} \operatorname{Im} \int_0^{2\pi} \frac{\sum_{n=1}^{N} c_n r_n \sin(\varphi - \varphi_n) \left(re^{i\varphi}\right)^m e^{-r_n^2 + 2rr_n \cos(\varphi-\varphi_n)}}{\sum_{m=1}^{N} c_n \left(re^{i\varphi}\right)^m e^{-r_n^2 + 2rr_n \cos(\varphi-\varphi_n)}} d\varphi \\
&= m - \frac{2r}{2\pi} \operatorname{Im} \int_0^{2\pi} \frac{\sum_{n=1}^{N} c_n r_n \sin(\varphi - \varphi_n) e^{-r_n^2 + 2rr_n \cos(\varphi-\varphi_n)}}{\sum_{m=1}^{N} c_n e^{-r_n^2 + 2rr_n \cos(\varphi-\varphi_n)}} d\varphi = m.
\end{aligned}
\tag{5.3}
$$

In (5.3) the imaginary part of the last integral is equal to zero, since it is real. It follows from (5.3) that the TC of the superposition of parallel identical single-ring LG beams with numbers $(0, m)$ is equal to m in the initial plane.

5.1.2 TC of the Superposition of Parallel Identical LG Beams with Different Weight Coefficients in the Far Field

We propose that there is a superposition of N identical one-ring LG beams displaced from the optical axis in the initial plane. Then the complex amplitude in the initial plane is equal to:

$$
E_m(x, y) = \sum_{n=1}^{N} c_n \left\{ \frac{\sqrt{2}}{w_0} \left[(x - a_n) + i(y - b_n) \right] \right\}^m \exp\left[-\frac{(x - a_n)^2 + (y - b_n)^2}{w_0^2} \right],
\tag{5.4}
$$

where (x, y) is Cartesian coordinates in the initial plane, w_0 is Gaussian beam waist, (a_n, b_n) are coordinates of beam centers, c_n are superposition coefficients. In contrast to (5.1), in (5.4) the superposition beams are taken with complex weight coefficients c_m and the Gaussian beam waist radius is explicitly distinguished.

In the far zone, the displacement of each LG beam turns out to be a slope of the wavefront, that is, in the far zone, the LG beams become axial, but with slopes. Therefore, the complex amplitude of the entire superposition in the far zone has the form:

$$E_m(r, \varphi, z >> z_0) = \exp\left(-\frac{r^2}{w_0^2}\right)\left(\frac{\sqrt{2}}{w_0} re^{i\varphi}\right)^m \times \sum_{n=1}^{N} c_n \exp(ika_n r \cos\varphi + ikb_n r \sin\varphi). \quad (5.5)$$

where (r, φ) are far-field polar coordinates. According to the formula of Berry [6], TC is equal to

$$TC = \frac{1}{2\pi} \lim_{r\to\infty} \operatorname{Im} \int_0^{2\pi} \sum_{n=1}^{N} c_n \frac{\frac{\partial}{\partial\varphi}\left[\exp(ika_n r \cos\varphi + ikb_n r \sin\varphi)\right]}{\sum_{n=1}^{N} c_n \exp(ika_n r \cos\varphi + ikb_n r \sin\varphi)} d\varphi + \frac{1}{2\pi} \lim_{r\to\infty} \operatorname{Im} \int_0^{2\pi} \frac{\frac{\partial}{\partial\varphi}\left[\left(re^{i\varphi}\right)^m\right]}{\left(re^{i\varphi}\right)^m} d\varphi. \quad (5.6)$$

Reducing the common factors in the numerator and denominator, we obtain further:

$$TC = m + \frac{1}{2\pi} \lim_{r\to\infty} \operatorname{Im} \int_0^{2\pi} \sum_{n=1}^{N} c_n \frac{\partial}{\partial\varphi} \left[\exp(ika_n r \cos\varphi + ikb_n r \sin\varphi)\right] \times \left[\sum_{n=1}^{N} c_n \exp(ika_n r \cos\varphi + ikb_n r \sin\varphi)\right]^{-1} d\varphi. \quad (5.7)$$

The second term is the TC of some additional field of the form (without a Gaussian envelope).

$$E_{\text{add}}(x, y) = \sum_{n=1}^{N} c_n \exp(ika_n x + ikb_n y). \quad (5.8)$$

Since the numbers a_n, b_n, c_n are arbitrary, formula (5.8) can describe a wide class of light fields. In particular, the additional field can be a vortex and therefore give an additional TC. For example, if we take $N = 4$, $c_1 = -i$, $c_2 = 1$, $c_3 = i$, $c_4 = -1$, $a_1 = -a_3 = r_0$, $a_2 = a_4 = 0$, $b_1 = b_3 = 0$, $b_2 = -b_4 = r_0$, then we get an additional field in the form

$$E_{\text{add}}(x, y) = 2\sin(kax) + 2i\sin(kay), \tag{5.9}$$

which is near the center approximately

$$E_{\text{add}}(x \approx 0, y \approx 0) = 2ka(x + iy), \tag{5.10}$$

that is, it contains a vortex of the first order. If all coefficients c_n are real in superposition (5.4), then we can show that

$$E^*_{\text{add}}(x, y) = E_{\text{add}}(-x, -y). \tag{5.11}$$

From (5.11) it follows that if there is a zero amplitude at some point $(x_{\text{null}}, y_{\text{null}})$, then the amplitude is also zero at the point $(-x_{\text{null}}, -y_{\text{null}})$, and near zero the amplitude is complex conjugate. That is, there is a "conjugate" vortex for each vortex in the field (5.8). The TCs of these "conjugate" vortices compensate each other, and therefore the TC of the field (5.8) with real coefficients c_m is equal to zero. Expression (5.11) is proved simply:

$$\begin{aligned} E^*(u, v) &= \sum_{n=1}^{N} c_n \exp(-ixu - iyv) \\ &= \sum_{n=1}^{N} c_n \exp(ix(-u) + iy(-v)) \\ &= E(-u, -v). \end{aligned} \tag{5.12}$$

There is also a physical reason why the field (5.8) cannot have other TC than zero. Indeed, the amplitude of the form (5.8) is formed in the Fourier plane (in the focus of a spherical lens) by a light field, which in the initial plane consists of N point sources with different amplitudes, but the same phase. A light field whose amplitude is a real function can only create OVs in pairs with $+p$ and $-p$ TCs. This also follows from the fact that an OV that has passed through the amplitude mask does not change its TC [17, 18].

If the TC of the superposition (5.1) in the initial plane and in the far zone is the same and equals m, then in any other plane it is equal to m, if the coefficients c_n are real.

5.1.3 Numerical Modeling

For example, Fig. 5.1 shows the intensities and phases of three superpositions of off-axis single-ring LG beams with the following parameters: the wavelength is λ

= 532 nm, the waist radius of all beams is $w_0 = 0.5$ mm, the number of LG beams is $N = 4$, the TC of each of them is $m = 3$, centers of these beams (in Cartesian coordinates) are $(a_1, b_1) = (r_0, 0)$, $(a_2, b_2) = (0, r_0)$, $(a_3, b_3) = (-r_0, 0)$, $(a_4, b_4) = (0, -r_0)$, where $r_0 = 3w_0$, the superposition coefficients for LG beams are $c_1 = c_2 = c_3 = c_4 = 1$ (Fig. 5.1a, d), $c_1 = -i$, $c_2 = 1$, $c_3 = i$, $c_4 = -1$ (Fig. 5.1b, e), and $c_1 = -i$, $c_2 = -1$, $c_3 = i$, $c_4 = 1$ (Fig. 5.1c, f, g), computational domain is restricted by $|x|, |y| \leq R$, where $R = 5$ mm (Fig. 5.1a–f) and $R = 10$ mm (Fig. 5.1g), the radius of the circle for calculating the TC is $R_1 = 4.5$ mm (Fig. 5.1d–f) and $R_1 = 9.5$ mm (Fig. 5.1g), the grid size in each direction is $N = 1024$. When all superposition coefficients are the same (Fig. 5.1a, d), the phase distribution becomes asymmetric, but the total TC of four LG beams turned out to be the same as for each of them: TC = 3.0042 ≈ 3. If we choose the coefficients, as in Fig. 5.1b, e, then, despite their dissimilarity, the phase distribution is symmetrical about the center and the total TC has changed and turned out to be equal to TC = 4.0003 ≈ 4. For the other coefficients (Fig. 5.1c, f, g), the total TC has also changed and is equal to TC = 5.9234 ≈ 6 (Fig. 5.1f). However, according to Eq. (5.2), the TC is computed over an infinite-radius circle and, thus, accounts all the vortices in the light field (or, equivalently, the 2π phase jumps in the beam periphery). Computation indicates that the figure size $2R = 10$ mm was sufficient to account all phase jumps in Fig. 5.1d, e, but insufficient for Fig. 5.1f. Figure 5.1g illustrates the phase distribution in a wider area ($2R = 20$ mm). In the periphery, there are four phase jumps by 2π and two phase jumps by -2π (denoted respectively by '+' and by '−'). Thus, the TC should be equal to $(4 \times 2\pi - 2 \times 2\pi)/2\pi = 2$. Numerical computation confirms it and yields the value TC = 1.9975 ≈ 2, which is, in contrast with Fig. 5.1b, e, less than the TC of the constituent LG beams. Thus, the superposition coefficients $c_1 = -i$, $c_2 = 1$, $c_3 = i$, $c_4 = -1$ (Fig. 5.1b, e) increment the TC of the whole superposition, whereas the coefficients $c_1 = -i$, $c_2 = -1$, $c_3 = i$, $c_4 = 1$ (Fig. 5.1c, f, g) decrement this TC.

When propagating in free space, four LG beams expand and begin to interfere with each other. Figure 5.2 shows the intensities and phases of the beams from Fig. 5.1 with the same parameters, but at the Rayleigh distance $z = z_0 = kw_0^2/2 \approx 1.476$ m. When all superposition coefficients are equal to each other (Fig. 5.2a, d), the total TC of four LG beams remains equal to three: $TC = 2.9968 \approx 3$. For the beam in Fig. 5.2b, e, the total TC remains equal to four: $TC = 3.9903 \approx 4$. For the beam in Fig. 5.2c, f, the total TC is equal to six ($TC = 5.9148 \approx 6$), but, again, choosing a wider domain (Fig. 5.2g) yields the value TC = 1.9713 ≈ 2.

In the far zone, all four LG beams mix with each other, and the distributions of their intensity and phase are shown in Fig. 5.3. All calculation parameters are the same as in Fig. 5.1, but the propagation distance $z = 3z_0 \approx 4.429$ m, computational area is limited by $|x|, |y| \leq R$, where $R = 7.5$ mm, circle radius for calculating TC is $R_1 = 7$ mm. When all superposition coefficients are the same (Fig. 5.3a, d), the total TC of four LG beams remains equal to three: $TC = 2.9875 \approx 3$. For the beam in Fig. 5.3b, e, the total TC is incremented and is equal to four: $TC = 3.9760 \approx 4$. For the beam in Fig. 5.3c, f, the total TC is decremented and is equal to two: $TC = 2.0036 \approx 2$.

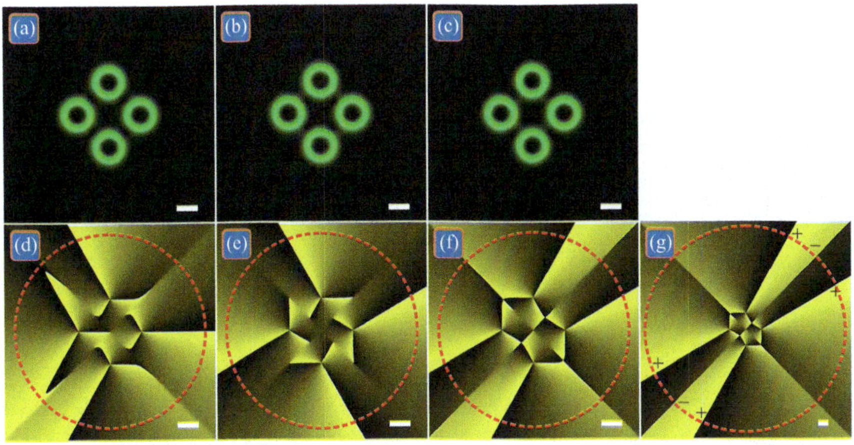

Fig. 5.1 Intensities and phases of three superpositions of off-axis single-ring LG beams, in which the total TC is the same as that of each beam **a**, **d**, or differs from the TC of each beam by +1 **b**, **e** and by −1 **c**, **f**, **g** due to the presence of complex weight coefficients in the superposition. Black and green intensity means respectively zero and maximal values. Black and yellow phase means respectively 0 and 2π. Here and in all other figures, red dashed circle on the phase distributions denotes the circle of the TC computation, whereas white scale marks in the right bottom denote 1 mm

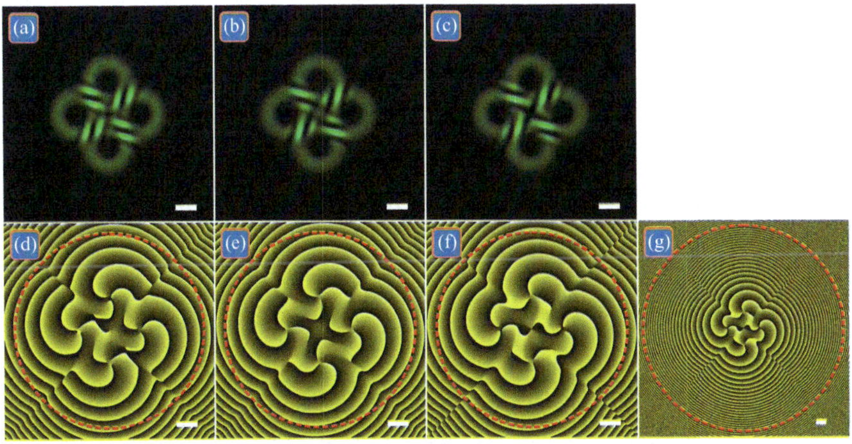

Fig. 5.2 Intensities and phases of three superpositions of off-axis single-ring LG beams with the initial distribution shown in Fig. 5.1 at Rayleigh distance

It should be noted that all LG beams in the superpositions shown in Figs. 5.1, 5.2 and 5.3 have the same power. Although, it follows from the theory above that if all superposition coefficients are real then the TC does not change even in the case of a superposition of LG beams with different power (LG beams are added in phase or in anti-phase). So, Fig. 5.4 shows two such superpositions. In one of them, the LG

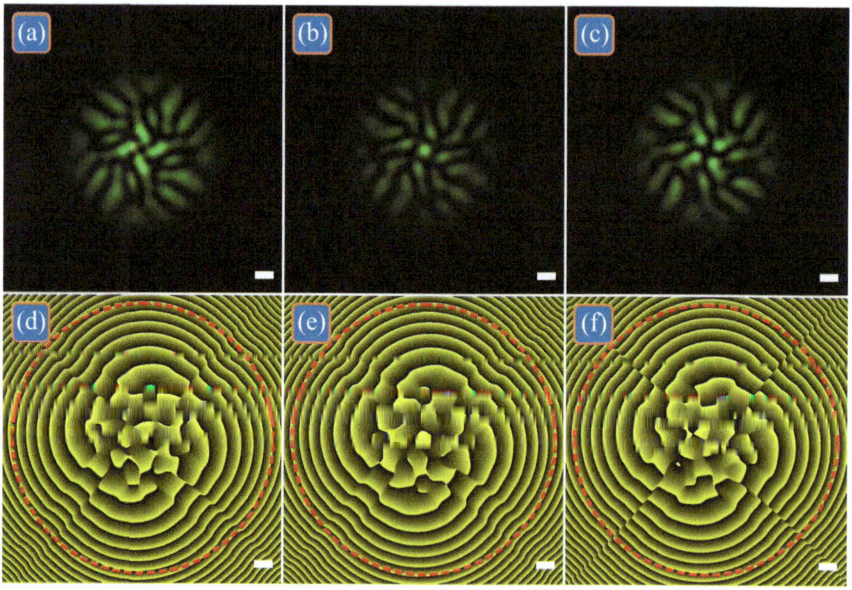

Fig. 5.3 Intensities and phases of three superpositions of off-axis single-ring LG beams with the initial distribution shown in Fig. 5.1 at triple Rayleigh distance (far zone)

beams on the horizontal axis are twice as powerful as the LG beams on the vertical axis: $c_1 = c_3 = 1$, $c_2 = c_4 = 1/2^{1/2}$ (Fig. 5.4a–d). In another superposition, the beam power decreases in a circle: $c_1 = 1$, $c_2 = 3^{1/2}/2$, $c_3 = 2^{1/2}/2$, $c_4 = 1/2$ (Fig. 5.4e–h). Other calculation parameters are the same as in Fig. 5.1, but the propagation distance $z = 0$ (Fig. 5.4a, b, e, f) and $z = z_0/2$ (Fig. 5.4c, d, g, h). The numerically calculated TC for both beams at both distances along the optical axis turned out to be equal to three: $TC = 3.0037$ (Fig. 5.4a, b), $TC = 2.9995$ (Fig. 5.4c, d), $TC = 3.0035$ (Fig. 5.4e, f), $TC = 2.9993$ (Fig. 5.4 g, h).

Thus, we have proved that the superposition of identical one-ring Laguerre-Gaussian beams with numbers (0, m) which are parallel to the optical axis, have the same phase, but different power in the initial plane, has a topological charge equal to m, regardless on the distance between the beams and on the power of each beam [19]. And only if the beams have a different phase in the initial plane (in the waist plane), then the superposition TC changes. The theoretical consideration is confirmed by the simulation results, which demonstrated how tuning the phase delays between the superposition coefficients allows the TC of the superposition to be conserved, or incremented, or decremented. The potential application areas are optical data transmission and doing simple arithmetic operations in optical computing machines where data are carried by vortex light beams and are encoded by the topological charges [20, 21].

In wireless optical data transmission, the results can be used by generating considered superpositions of LG beams using a SLM and by identifying the incoming

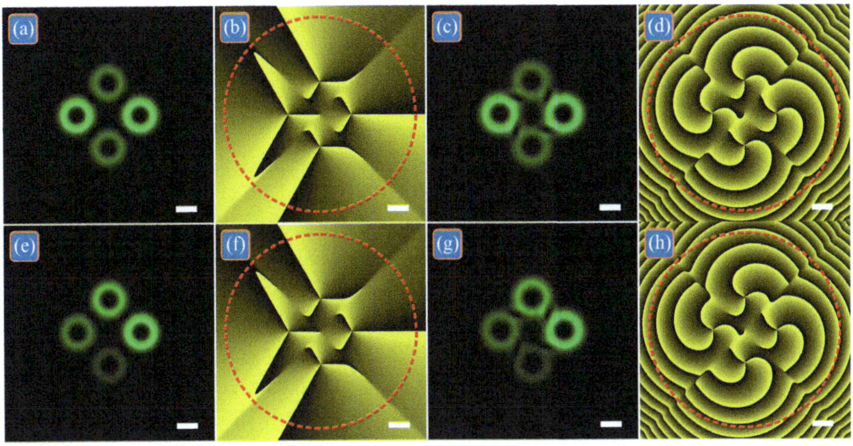

Fig. 5.4 Intensities and phases of two super positions of off-axis single-ring LG beams with different powers the initial plane and at the half of the Rayleigh distance

signals by the superposition's topological charge using a Shack-Hartmann wavefront sensor. According to Fig. 5.3, simple adding phase delays between the LG beams allows changing the TC of the whole superposition, whereas using the superposition of the LG beams instead of separate LG beams increases the resistance to the turbulence-induced distortions, adds degrees-of-freedoms in data encoding, and improves the data security, since the TC of a single LG mode can be determined simply by the radius of the light ring, while the TC of the superposition depends not only on the radii of the constituent beams, but also on the phase delays between them.

In optical computing, for designing compact devices, the LG beams should be localized in guiding microstructures rather than propagating in free space. However, implementation of LG beams propagation in on-chip devices is now challenging for typical semiconductor manufacturing processes since most of integrated optical waveguides do not have circular symmetry in their transverse geometries in contrast to optical fibers. However, such techniques are developed now. For instance, a ± 1st-order vortex can be implemented in a rectangular-shaped waveguide as a superposition of the TE01 and TE10 modes with a phase delay of $\pi/2$ [22]. In [23], an integrated cross-shaped waveguide structure to support high-order OAM modes up to 4th order was proposed. This waveguide was designed for the wavelength of 1550 nm and has the transverse sizes just 1.626 μm × 1.504 μm. The obtained here results show that simple free-space mixing of the LG beams, outcoming from such structures, can generate light fields with incremented or decremented TC, although we did not investigate the influence of the mode purity of the LG beams.

5.2 Geometric Progression of Optical Vortices

Optical vortices are actively studied nowadays. For example, in [24], cylindrical vector spatiotemporal optical vortices were experimentally studied. In [25], an optical setup was proposed for generating a 2D optical vortex lattice. In [26], a 3D optical vortex lattice was generated experimentally. The work [27] contains a review about using optical vortices for trapping and rotating microscopic particles. The paper [28] investigates optical vortices in nanophotonics. Optical vortices in structured waveguides are studied in [29]. In [30], distortions of spiral vortex beams are considered and it is demonstrated that such beams are almost insensitive to the distortions.

For most well known optical vortices, the orbital angular momentum (OAM) (normalized to the beam power) and the topological charge (TC) have the same value. This is the case for the Laguerre-Gaussian beams [5], Bessel modes [31] and Bessel-Gaussian beams [32], Gaussian optical vortices [33], hypergeometric modes [34] and beams [35], circular beams [36], elliptical beams [37], and many other well-known vortex beams. As to the beams without cylindrical symmetry, for instance, asymmetric Bessel beams [38] or Hermite-Gaussian vortex modes [39], their OAM and TC typically have different values.

Coaxial superpositions of optical vortices can also have different values [18] of the TC and of the normalized OAM. The OAM can have both integer and fractional value [18] and it conserves on propagation. If the TC in the initial plane is fractional, then, on propagation, it acquires an integer value, though indefinite [40] since it is 'unclear' how the fractional number should be rounded: to either greater or lower nearest integer. In addition, if a single Gaussian optical vortex has a half-integer TC in the initial plane, for instance, $n + 1/2$, then, on propagation, the TC becomes integer, but undefined since at different distances from the optical axis the TC is either n or $n + 1$. In the case of a coaxial superposition of two Gaussian optical vortices, if the TC in the initial plane is half-integer ($(n + m)/2$), then, on propagation of such vortex field, the TC becomes integer and equal to $\max(n, m)$, and conserves on further propagation [18].

In this section, we investigate several variants of coaxial superpositions of Gaussian optical vortices, whose complex amplitude has the form of a geometric progression (either finite or infinite) or of a Newton's binomial. It is shown that the TC of such superpositions in the initial plane can be either integer or half-integer. On propagation in free space, the TC can be only integer and conserves its value. In the general case, the geometric progression of optical vortices (GPOV) is a four-parameter family of laser beams, three parameters are integer (k, n, m) and one parameter is real (a). The topological charge of the GPOV in the initial plane depends on all four parameters, but after propagation in free space, the TC is equal to the TC of the constituent Gaussian optical vortex with the larger weight coefficient in the superposition.

5.2.1 Geometric Progression of Optical Vortices in the Initial Plane

Here we consider a coaxial superposition of the Gaussian optical vortices that can be described by a geometric progression. In the initial plane, such a superposition has the following complex amplitude:

$$\begin{aligned} E_{1n}(r,z) &= \exp\left(-\frac{r^2}{w^2}\right)\left(1+e^{i\varphi}+e^{i2\varphi}+\cdots+e^{in\varphi}\right) \\ &= \exp\left(-\frac{r^2}{w^2}\right)\left(\frac{e^{i(n+1)\varphi}-1}{e^{i\varphi}-1}\right). \end{aligned} \tag{5.13}$$

Equation (5.13) can be converted so that the argument of the complex number [phase of the field (5.13)] would be written explicitly:

$$\begin{aligned} E_n(r,z) &= \exp\left(-\frac{r^2}{w^2}\right)\left(1+e^{i\varphi}+e^{i2\varphi}+\cdots+e^{in\varphi}\right) \\ &= \exp\left(-\frac{r^2}{w^2}+\frac{in\varphi}{2}\right)\frac{\sin\left(\frac{n+1}{2}\right)\varphi}{\sin\left(\frac{\varphi}{2}\right)}. \end{aligned} \tag{5.14}$$

It is seen in Eq. (5.14) that the argument (phase) of the superposition (5.13) is equal to $(n\varphi)/2$. Thus, the topological charge TC of the vortex field (5.13) is

$$TC = n/2. \tag{5.15}$$

It is known that the orbital angular momentum (OAM) J_z of a paraxial beam, normalized to its power W, reads as [18]

$$J_z = \mathrm{Im}\int_0^\infty\int_0^{2\pi} E^*(r,\varphi)\frac{\partial E(r,\varphi)}{\partial\varphi}r\,dr\,d\varphi, \tag{5.16}$$

$$W = \int_0^\infty\int_0^{2\pi} E^*(r,\varphi)E(r,\varphi)r\,dr\,d\varphi, \tag{5.17}$$

with $E^*(x, y)$ being the complex conjugate amplitude. It can be shown that the normalized OAM of the beam (5.14) is also equal to (5.15), i.e. $J_z/W = n/2$.

It is also seen from Eq. (5.14) that the polynomial $1+z+z^2+z^3+\cdots+z^n$ (with $z=e^{i\varphi}$) has n roots. Thus, the field (5.14) has n zero-amplitude rays outgoing from the center at the angles $\varphi = 2\pi p/(n+1)$ ($p = 1, \ldots, n$). In the topological sense, these rays are edge dislocations, since the phase along a contour around the center jumps by π when intersecting these zero-intensity rays. The intensity distribution of the field (5.14) has a shape of a light lobe elongated in the positive direction of the horizontal axis, and the maximal intensity is on the ray $\varphi = 0$. This follows from resolving the 0/0 uncertainty in Eq. (5.14) at $\varphi = 0$. Maximal intensity is equal to $(n + 1)^2$ at $r = 0$. In addition, the field (5.14) has $(n - 1)$ intensity side lobes residing between n edge dislocations. Thus, the total number of intensity lobes (central and side lobes) is equal to n.

5.2.2 *Geometric Progression of Optical Vortices in the Fresnel Diffraction Zone*

To derive the TC on propagation of the field (5.14), it is necessary to obtain an asymptotic of the amplitude of the field (5.14) at arbitrary $z > 0$ and at $r \to \infty$. At first, we apply the Fresnel transform to obtain the amplitude of each term in Eq. (5.14). If in the initial plane there is a Gaussian optical vortex

$$E_n(r, \varphi) = e^{-r^2/w^2 + in\varphi}, \tag{5.18}$$

then at a distance z from the initial plane its complex amplitude is given by [18]:

$$E_n(\rho, \theta, z) = (-i)^{n+1}\sqrt{\frac{\pi}{2}\frac{z_0}{zq}}\exp\left(\frac{ik\rho^2}{2z} + in\theta\right) \times \sqrt{\xi}\exp(-\xi)\left[I_{\frac{n-1}{2}}(\xi) - I_{\frac{n+1}{2}}(\xi)\right], \tag{5.19}$$

with

$$\xi = \left(\frac{z_0}{z}\right)^2\left(\frac{\rho}{w}\right)^2\left(\frac{1}{2q}\right), \quad q = 1 - i\frac{z_0}{z}. \tag{5.20}$$

In Eq. (5.19), $I_\mu(x)$ is the modified Bessel function, k is the wavenumber of light, $z_0 = kw^2/2$ is the Rayleigh distance, (ρ, θ) are the polar coordinates in the transverse plane at the distance z. For large values of the argument $\rho >> w$, an asymptotic expansion of the modified Bessel function can be adopted with only two terms. Thus, an approximate expression is derived for the difference of two modified Bessel functions of the neighbor orders at large values of the argument:

$$I_{\frac{n-1}{2}}(\xi) - I_{\frac{n+1}{2}}(\xi) \sim \frac{e^\xi}{\sqrt{2\pi\xi}}\left\{1 - \frac{1}{8\xi}\left[4\left(\frac{n-1}{2}\right)^2 - 1\right]\right\} - \frac{e^\xi}{\sqrt{2\pi\xi}}\left\{1 - \frac{1}{8\xi}\left[4\left(\frac{n+1}{2}\right)^2 - 1\right]\right\} = \frac{ne^\xi}{2\xi\sqrt{2\pi\xi}}. \tag{5.21}$$

Then, the field (5.14) at large values $\rho >> w$ is given by

$$E_n(\rho, \theta, z) = \frac{-iz_0}{4zq\xi} \exp\left(\frac{ik\rho^2}{2z}\right) \sum_{p=1}^{n} p \exp\left(ip\overline{\theta}\right)$$
$$= \frac{-iz_0}{4zq\xi} \exp\left(\frac{ik\rho^2}{2z}\right) \left[\frac{e^{i\overline{\theta}}\left(1 + ne^{i(n+1)\overline{\theta}} - (n+1)e^{in\overline{\theta}}\right)}{\left(1 - e^{i\overline{\theta}}\right)^2}\right], \tag{5.22}$$

with $\overline{\theta} = \theta - \pi/2$. To obtain the expression (5.22), we used that if some function is of the form

$$F(\varphi) = \sum_{p=0}^{n} e^{ip\varphi},$$

then the expression

$$\sum_{p=0}^{n} pe^{ip\varphi}$$

is equal to the derivative $-i\partial F(\varphi)/\partial\varphi$. It can be shown that at $\overline{\theta} = 0$ (i.e. $\theta = \pi/2$), the intensity is maximal. The expression in square brackets in Eq. (5.22) can be rewritten as

$$\frac{e^{i\overline{\theta}}\left(1 + ne^{i(n+1)\overline{\theta}} - (n+1)e^{in\overline{\theta}}\right)}{\left(1 - e^{i\overline{\theta}}\right)^2} = -\frac{\left(1 + ne^{i(n+1)\overline{\theta}} - (n+1)e^{in\overline{\theta}}\right)}{4\sin^2\left(\frac{\overline{\theta}}{2}\right)}. \tag{5.23}$$

At large values of n, we can neglect the unity in the numerator in Eq. (5.23) and, therefore, the TC of the field (5.22) is defined by the competition of two optical vortices:

$$ne^{i(n+1)\overline{\theta}} - (n+1)e^{in\overline{\theta}}. \tag{5.24}$$

As was shown in [18], the TC of a superposition of two optical vortices, similar to Eq. (5.24), equals the TC of a vortex with a greater amplitude. Since $n + 1 > n$, the TC of the field (5.14) at a distance z from the initial plane is equal to $TC = n$. This is explainable since in the superposition (5.22) $\Sigma pe^{ip\varphi}$ ($p = 1, \ldots, n$), the last term $n\exp(in\varphi)$ has the maximal coefficient (beam power) and therefore the TC of this term 'wins' in the topological competition. From Eq. (5.24) follows that despite the TC of the field (5.14) in the initial plane equals $TC = n/2$, further, on propagation of the field (5.14) in free space, the TC is equal to $TC = n$.

5.2.3 Truncated Geometric Progression of Optical Vortices

Now we revisit the field (5.14) and consider its modifications. It is interesting that removing the vortex-free term (i.e. 1) from the amplitude of the light field (5.13) leads to the following expression, instead of Eq. (5.14):

$$\begin{aligned} E_{2n}(r,z) &= \exp\left(-\frac{r^2}{w^2}\right)\left(e^{i\varphi}+e^{i2\varphi}+\cdots+e^{in\varphi}\right) \\ &= \exp\left(-\frac{r^2}{w^2}+\frac{i(n+1)\varphi}{2}\right)\frac{\sin\left(\frac{n\varphi}{2}\right)}{\sin\left(\frac{\varphi}{2}\right)}. \end{aligned} \tag{5.25}$$

According to Eq. (5.25), the topological charge of the field (5.25) equals

$$TC=(n+1)/2. \tag{5.26}$$

When the light field propagates ($z > 0$), asymptotic of its complex amplitude at large distances from the optical axis ($r \to \infty$) is described by an expression exactly the same as Eq. (5.22). Thus, TC of the field (5.25) at $z > 0$ is the same as that of the field (5.13), i.e. equal to $TC = n$.

Similarly, instead of the field (5.25), we can consider a field described by the geometric progression starting with kth term. So, removing the first ($k - 1$) terms from Eq. (5.25), we get:

$$\begin{aligned} E_{kn}(r,z) &= \exp\left(-\frac{r^2}{w^2}\right)\left(e^{ik\varphi}+e^{i(k+1)\varphi}+\cdots+e^{in\varphi}\right) \\ &= \exp\left(-\frac{r^2}{w^2}+\frac{i(n+k)\varphi}{2}\right)\frac{\sin\left(\frac{(n-k)\varphi}{2}\right)}{\sin\left(\frac{\varphi}{2}\right)}. \end{aligned} \tag{5.27}$$

It is seen that the field (5.27) has the phase $(n + k)\varphi/2$. Therefore, the TC is equal to $TC = (n + k)/2$. It is interesting that if only one term is remained in the field (1) with the order $k = n$, then the formula (5.27) yields the integer TC equal to $TC = n$.

5.2.4 Geometrical Progression of Optical Vortices with a Symmetric OAM-Spectrum

We note that the geometric progressions (5.13), (5.25) and (5.27) have a uniform OAM-spectrum, that is, all the angular harmonics in these superpositions have the same coefficients equal to unity. As was shown in [41], if the OAM-spectrum of a light field is symmetric, then its normalized OAM is equal to the order (topological charge) of the central angular harmonic. This is fully applicable to the fields (5.13),

(5.25) and (5.27). Indeed, the average order of the angular harmonic in the center of the OAM-spectrum is $n/2$, $(n + 1)/2$ and $(n + \mathrm{k})/2$, respectively. It can be proved that if the superposition (5.13) has real coefficients and a symmetric OAM-spectrum, then the TC in the initial plane of such a superposition is equal to the order of the central harmonic. Indeed, instead of the progression (5.13) we consider a superposition of the form

$$E_{sn}(r, z) = \exp\left(-\frac{r^2}{w^2}\right)\left(C_0 + C_1 e^{i\varphi} + C_2 e^{i2\varphi} + \cdots + C_n e^{in\varphi}\right). \quad (5.28)$$

We suppose that the coefficients are symmetric with respect to the central coefficient with the number $n_0 = n/2$, i.e. $C_{n_0-k} = C_{n_0+k}$. If n is an odd number, then the center of the OAM-spectrum of the field (5.28) resides in the middle between the orders $(n - 1)/2$ and $(n + 1)/2$. For such symmetrical coefficients, the field (5.28) can be written as

$$\begin{aligned} E_{sn}(r, z) &= \exp\left(-\frac{r^2}{w^2}\right)\sum_{k=0}^{n} C_k e^{ik\varphi} \\ &= \exp\left(-\frac{r^2}{w^2}\right) \\ &\quad \left(C_{n_0} e^{in_0\varphi} + \sum_{k=1}^{n_0}\left(C_{n_0-k} e^{i(n_0-k)\varphi} + C_{n_0+k} e^{i(n_0+k)\varphi}\right)\right) \\ &= \exp\left(-\frac{r^2}{w^2}\right)\left(C_{n_0} e^{in_0\varphi} + 2e^{in_0\varphi}\sum_{k=1}^{n_0} C_{n_0-k}\cos k\varphi\right) \\ &= \exp\left(-\frac{r^2}{w^2} + in_0\varphi\right)\left(C_{n_0} + 2\sum_{k=1}^{n_0} C_{n_0-k}\cos k\varphi\right). \end{aligned} \quad (5.29)$$

Since the expression in the right brackets is real-valued, the TC of the expression (5.29) is equal to the order of the central angular harmonic $TC = n_0 = n/2$. As an example of a superposition (5.28) with the symmetric OAM-spectrum, we can consider the following field ($a > 0$):

$$\begin{aligned} E_{sn}(r, z) &= \exp\left(-\frac{r^2}{w^2}\right)\sum_{k=-n_0}^{n_0}\frac{a^{|k|}}{|k|!}\exp(i(n_0 + k)\varphi) \\ &= 2\exp\left(-\frac{r^2}{w^2} + in_0\varphi\right)\sum_{k=0}^{n_0}\frac{a^{|k|}}{|k|!}\cos(k\varphi), \\ n_0 &= n/2. \end{aligned} \quad (5.30)$$

Another example of a superposition with the symmetric OAM-spectrum is a superposition of optical vortices with their weight coefficients chosen so as to describe the Newton's binomial:

$$E_{3n}(r,z) = \exp\left(-\frac{r^2}{w^2}\right)\left(1+e^{i\varphi}\right)^n$$
$$= 2^n \exp\left(-\frac{r^2}{w^2}+\frac{in\varphi}{2}\right)\left(\cos\left(\frac{\varphi}{2}\right)\right)^n. \quad (5.31)$$

Equation (5.31) indicates that the topological charge of the superposition of optical vortices in the form of the Newton's binomial is also integer or half-integer and is equal to the expression (5.15): $TC = n/2$.

Instead of Eq. (5.31), a superposition of two optical vortices can be studied, with their amplitude raised to a power:

$$E_{3n}(r,z) = \exp\left(-\frac{r^2}{w^2}\right)\left(e^{ip\varphi}+e^{iq\varphi}\right)^n$$
$$= 2^n \exp\left(-\frac{r^2}{w^2}+\frac{in(p+q)\varphi}{2}\right)$$
$$\left(\cos\left(\frac{\varphi(p-q)}{2}\right)\right)^n. \quad (5.32)$$

The topological charge of the superposition of optical vortices (5.32) is equal to

$$TC = n(p+q)/2 \quad (5.33)$$

and can be integer or half-integer.

5.2.5 *Unbounded Geometric Progression of Optical Vortices*

It is interesting to know the TC value of the light field (5.13) if it is written as an unbounded geometric progression:

$$E_{\infty}(r,z) = \exp\left(-\frac{r^2}{w^2}\right)$$
$$\left(1+qe^{i\varphi}+q^2e^{i2\varphi}+\cdots+q^ne^{in\varphi}+\ldots\right)$$
$$= \exp\left(-\frac{r^2}{w^2}\right)\left(\frac{1}{1-qe^{i\varphi}}\right), \quad |q|<1, \quad (5.34)$$

The argument (phase) of the complex amplitude (5.34) reads as

$$\arg E_{\infty}(r,z) = \arctan\left(\frac{q\sin\varphi}{1-q\cos\varphi}\right). \quad (5.35)$$

Substituting the function (5.35) into the Berry's formula [6] for the TC calculation, we get

$$TC = \frac{1}{2\pi}\lim_{r\to\infty}\left\{\int_0^{2\pi}\frac{\partial \arg E_\infty(r,\varphi)}{\partial\varphi}d\varphi\right\}$$
$$= \frac{1}{2\pi}\int_0^{2\pi}\frac{q\cos\varphi - q^2}{1+q^2-2q\cos\varphi}d\varphi = 0,\quad |q|<1. \tag{5.36}$$

Equation (5.36) indicates that despite the field (5.34) is a superposition of an infinite number of optical vortices, its TC at $|q| < 1$ is equal to zero.

Similarly, it can be shown that the infinite geometric progression obtained on the base of the field (5.35)

$$E_{1\infty}(r,z) = \exp\left(-\frac{r^2}{w^2}\right)\left(qe^{i\varphi} + q^2e^{i2\varphi} + \cdots + q^ne^{in\varphi} + \cdots\right)$$
$$= \exp\left(-\frac{r^2}{w^2}\right)\left(\frac{qe^{i\varphi}}{1-qe^{i\varphi}}\right),\quad |q|<1, \tag{5.37}$$

has the TC equal to 1. Both these results are explainable since the TC of the superposition (5.34) or (5.37) equals to the TC of the constituent vortex with the maximal weight coefficient (power). In the superposition (5.34), the first term (equal to 1) has the maximal partial power, and thus the TC equals zero. In the superposition (5.37), maximal power is also of the first term $q\exp(i\varphi)$ and, therefore, TC is equal to 1.

5.2.6 Superposition of Optical Vortices Described by a Geometric Progression with the Common Ratio

Here we study a superposition of optical vortices described by a geometric progression with a common ratio. Such a superposition has the following complex amplitude in the initial plane:

$$E_{1nk} \tag{5.38}$$

If the weight coefficients are of the same value by modulus (i.e. $|a| = 1$), the expression (5.38) can be converted into a form with explicit argument of a complex number [phase of the field (5.13)]:

$$E_{1nk}(r,\varphi)=\exp\left(-\frac{r^2}{w^2}+i\frac{k+n}{2}m\varphi\prime\right)\frac{\sin\left(\frac{n-k+1}{2}m\varphi\prime\right)}{\sin\left(\frac{1}{2}m\varphi\prime\right)},\tag{5.39}$$

with φ' = φ + m^{-1} arg a. As seen from Eq. (5.39), the argument (phase) of the superposition (5.13) is equal to [$m(n + k)$φ']/2. Thus, the topological charge TC of the vortex field (5.13) is equal to

$$TC=m(n+k)/2.\tag{5.40}$$

If $m = n = 1$, the TC (5.40) coincides with Eq. (5.26). However, in a general case, when $|a| \neq 1$, the TC can be obtained by using the residues theory. Substituting the right part of Eq. (5.38) into the M.V. Berry's formula (5.36), we get:

$$\begin{aligned}TC&=\frac{1}{2\pi}\lim_{r\to\infty}\mathrm{Im}\int_0^{2\pi}d\varphi\frac{\partial E(r,\varphi)/\partial\varphi}{E(r,\varphi)}\\&=\frac{1}{2\pi}\mathrm{Im}\int_0^{2\pi}\left[\frac{imae^{im\varphi}}{1-ae^{im\varphi}}+\frac{ikm-i(n+1)ma^{n-k+1}e^{i(n-k+1)m\varphi}}{1-a^{n-k+1}e^{i(n-k+1)m\varphi}}\right]d\varphi\\&=mn+\frac{m}{2\pi}\mathrm{Re}\int_0^{2\pi}\frac{d\varphi}{1-ae^{im\varphi}}-\frac{m(n-k+1)}{2\pi}\\&\quad\mathrm{Re}\int_0^{2\pi}\frac{d\varphi}{1-a^{n-k+1}e^{i(n-k+1)m\varphi}}.\end{aligned}\tag{5.41}$$

The first integral in Eq. (5.41) is equal to

$$\int_0^{2\pi}\frac{d\varphi}{1-ae^{im\varphi}}=\begin{cases}2\pi, & |a|<1,\\ \pi, & |a|=1,\\ 0, & |a|>1.\end{cases}\tag{5.42}$$

The second integral is evaluated similarly and equals the first one. Thus, we finally obtain that the TC of the geometric progression of optical vortices (5.48) in the initial plane significantly depends on the parameter a and is equal to

$$TC=\begin{cases}km, & |a|<1,\\ \frac{k+n}{2}m, & |a|=1,\\ nm, & |a|>1.\end{cases}\tag{5.43}$$

5.2.7 Numerical Simulation

Since the superposition (5.38) generalizes the superpositions (5.13), (5.25) and (5.27), the simulation is based on Eq. (5.38). Figure 5.5 depicts the intensity and phase distributions of the light field (5.38) at $n = 3$, $k = 0$, $m = 1$, $a = 1$, $w_0 = 500$ µm in different transverse planes: $z = 0$ (Fig. 5.5a, i), $z = z_0/200$ (Fig. 5.5b, j), $z = z_0/50$ (Fig. 5.5c, k), $z = z_0/20$ (Fig. 5.5d, l), $z = z_0/10$ (Fig. 5.5e, m), $z = z_0/2$ (Fig. 5.5f, n), $z = z_0$ (Fig. 5.5 g, o), $z = 2z_0$ (Fig. 5.5 h, p).

Figure 5.5a confirms that the intensity distribution in the initial plane has a shape of a light lobe ($m = 1$) elongated along the positive direction of the horizontal axis. However, the divergence of optical vortices increases with their topological charge.

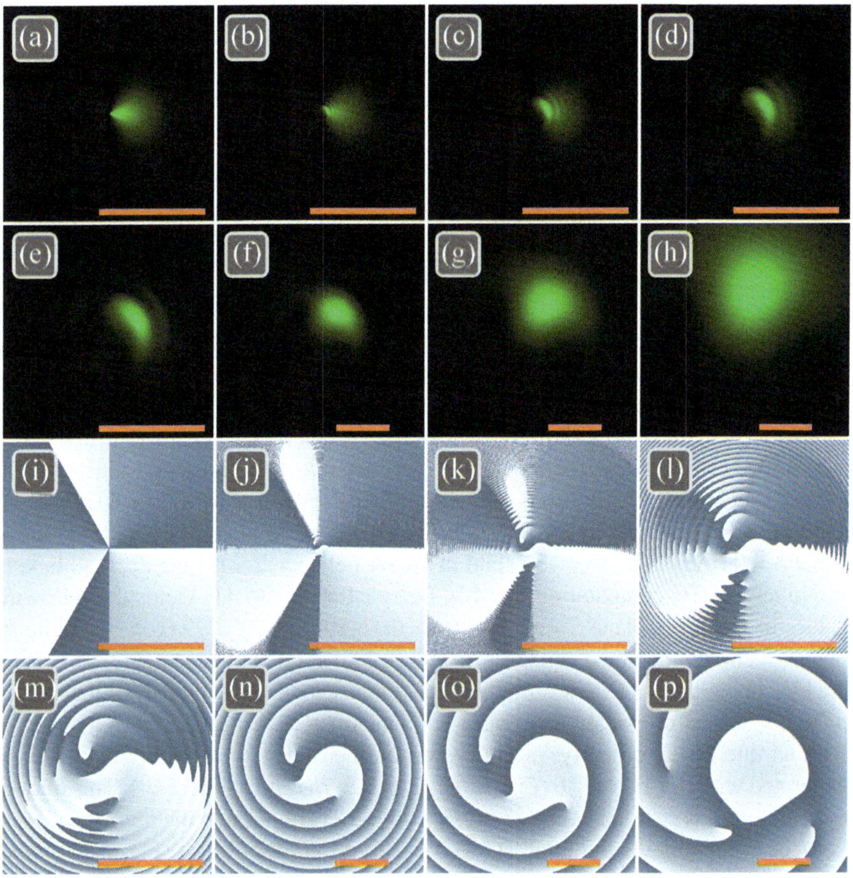

Fig. 5.5 Intensity (**a**–**h**) and phase (**i**–**p**) distributions of the light field (5.13) at $n = 3$, $k = 0$, $m = 1$, $a = 1$, $w_0 = 500$ µm in the planes $z = 0$ (**a**, **i**), $z = z_0/200$ (**b**, **j**), $z = z_0/50$ (**c**, **k**), $z = z_0/20$ (**d**, **l**), $z = z_0/10$ (**e**, **m**), $z = z_0/2$ (**f**, **n**), $z = z_0$ (**g**, **o**), $z = 2z_0$ (**h**, **p**). Scale mark in each figure means 1 mm

Therefore, all the constituent vortices in the superposition (5.13) diverge differently, and on propagation, the lobe splits into light bows, as can be seen in Fig. 5.5b–d. Then, on further propagation, since the higher divergence of vortices with a large topological charge leads to their stronger attenuation, only one light spot remains in the picture (Fig. 5.5f–h).

Numerical TC computation by M.V. Berry's formula (5.41) yields the following values: 1.49 ($z = 0$), 0.98 ($z = z_0/200$), 2.29 ($z = z_0/50$), 2.97 ($z = z_0/20$), 2.97 ($z = z_0/10$), 2.99 ($z = z_0/2$), 3.00 ($z = z_0$), 3.00 ($z = 2z_0$). Thus, the half-integer TC of $n/2$ in the initial plane ($TC = 1.5$) acquires an integer value n ($TC = 3$) on propagation in space.

To demonstrate the influence of the topological charge increment m in the progression, Fig. 5.6 depicts the intensity and phase distributions of the light field (5.38) at $n = 3$, $k = 0$, $m = 5$, $a = 1$, $w_0 = 500$ μm in different transverse planes: $z = 0$ (Fig. 5.6a, i), $z = z_0/200$ (Fig. 5.6b, j), $z = z_0/50$ (Fig. 5.6c, k), $z = z_0/20$ (Fig. 5.6d, l), $z = z_0/10$ (Fig. 5.6e, m), $z = z_0/2$ (Fig. 5.6f, n), $z = z_0$ (Fig. 5.6 g, o), $z = 2z_0$ (Fig. 5.6 h, p).

Figure 5.6a confirms that the intensity distribution in the initial plane consists of five lobes ($m = 5$) elongated at angles $\varphi = 2\pi p/5$ ($p = 0, \ldots, 4$). As in the case with $m = 1$, due to the different divergence of optical vortices on propagation, the lobes first split into the light bows (Fig. 5.6b–e) and then only one spot remains in the pattern with the intensity maximum on the optical axis (Fig. 5.6f–h).

Numerical TC computation based on M.V. Berry's formula (5.41) gives the following values: 7.48 ($z = 0$), −1.37 ($z = z_0/200$), 12.84 ($z = z_0/50$), 14.64 ($z = z_0/20$), 14.64 ($z = z_0/10$), 14.98 ($z = z_0/2$), 14.94 ($z = z_0$), 14.77 ($z = 2z_0$). Thus, half-integer TC of $mn/2$ ($TC = 7.5$) in the initial plane acquires the integer value mn ($TC = 15$) on propagation in space.

Further we study how the topological charge changes when several first terms are removed from the progression (5.38), i.e. when $k \neq 0$. Figure 5.7 illustrates the intensity and phase distributions of the light field (5.38) at $n = 11$, $k = 0$ (Fig. 5.7a–d), $k = 2$ (Fig. 5.7e–h), $k = 5$ (Fig. 5.7i–l), $m = 1$, $a = 1$, $w_0 = 500$ μm in the initial plane $z = 0$ (Fig. 5.7a, b, e, f, i, j) and in the far field $z = 2z_0$ (Fig. 5.7c, d, g, h, k, l).

Numerical TC computation by M.V. Berry's formula (5.41) yields the following values in the initial plane $z = 0$: 5.49 ($k = 0$), 6.49 ($k = 2$), 7.99 ($k = 5$). In the far field $z = 2z_0$, the obtained values are 10.96 (for all k, i.e. $k = 0$, $k = 2$, $k = 5$). Thus, the half-integer TC of $(n + k)/2$ in the initial plane becomes equal to the integer value n ($TC = 11$) on propagation in space. Optical vortices in Fig. 5.7 have a structure with an interesting property. In the center, on the optical axis, an optical vortex is generated with the TC of k: 0 (Fig. 5.7d), 2 (Fig. 5.7h) and 5 (Fig. 5.7l). This on-axis optical vortex is surrounded by vortices with the TC of +1 and the number of these vortices complements the total TC to $n = 11$: 11 (Fig. 5.7d), 9 (Fig. 5.3h) and 6 (Fig. 5.7l).

In this section, we derived the topological charge (TC) of a four-parametric family of vortex beams whose complex amplitude is described by a geometric progression of the Gaussian optical vortices [42]. This progression can be either raising, when the amplitudes of the constituent vortices increase, or decaying, or stationary. The

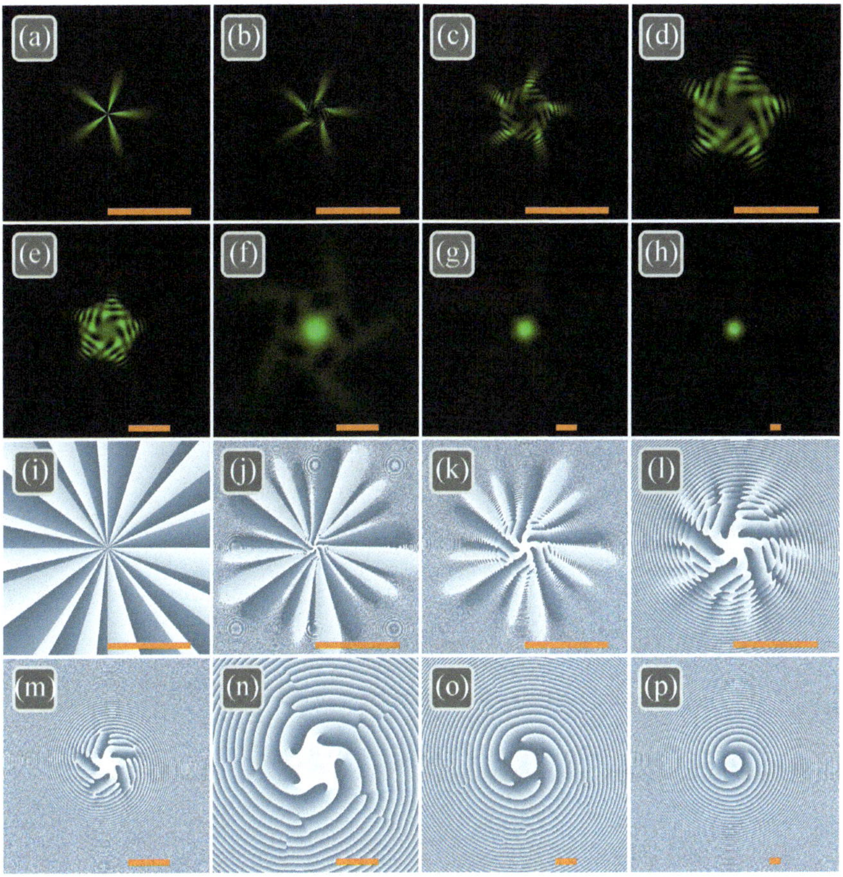

Fig. 5.6 Intensity (**a**–**h**) and phase (**i**–**p**) distributions of the light field (5.13) at $n = 3$, $k = 0$, $m = 5$, $a = 1$, $w_0 = 500$ μm in the planes $z = 0$ (**a**, **i**), $z = z_0/200$ (**b**, **j**), $z = z_0/50$ (**c**, **k**), $z = z_0/20$ (**d**, **l**), $z = z_0/10$ (**e**, **m**), $z = z_0/2$ (**f**, **n**), $z = z_0$ (**g**, **o**), $z = 2z_0$ (**h**, **p**). Scale mark in each figure means 1 mm

first and the last terms in this progression are equal respectively to $a^k \exp(-(r/w)^2 + ikm\varphi)$ and $a^n \exp(-(r/w)^2 + inm\varphi)$, i.e. the common ratio is $a \exp(im\varphi)$. The studied vortex beams family is described by three integer parameters (k, n, m) and by one real parameter a. If $a < 1$, the progression is decaying and the TC of the whole superposition is equal to the TC of its first term ($TC = km$, $k < n$), since this term describes an optical vortex of the maximal power in the superposition. If $a > 1$, the progression is raising and its TC is equal to the TC of the last term ($TC = nm$), since it is this term that has the maximal power in the superposition. Finally, if $a = 1$, the progression is stationary, its OAM-spectrum is symmetric and the TC of the whole superposition is equal to the order of the average angular harmonic ($TC = (k + n)m/2$). In the latter case, the superposition can have a half-integer TC in the initial plane. On propagation in free space though, the TC of the stationary progression of optical

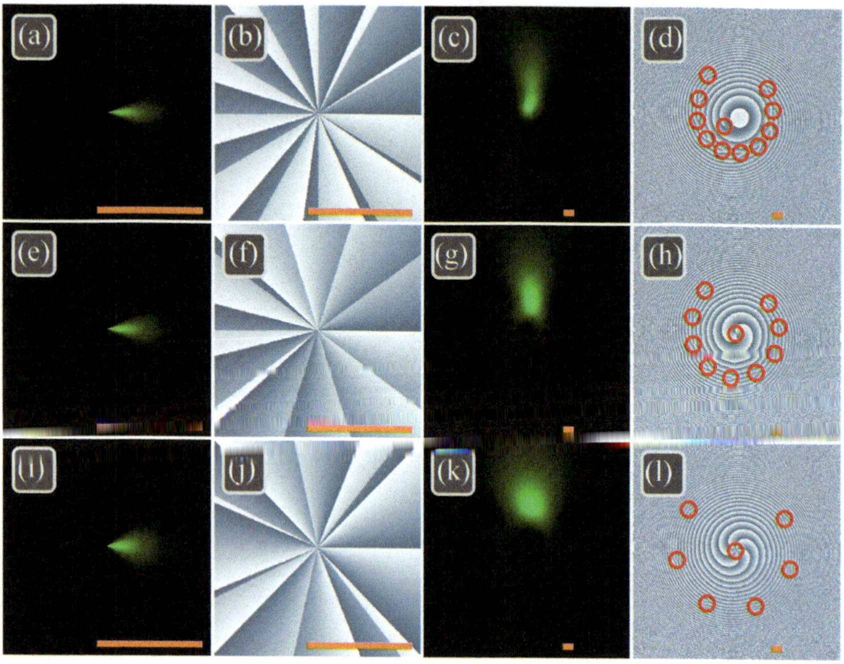

Fig. 5.7 Intensity (columns 1 and 3) and phase (columns 2 and 4) distributions of the light field (5.13) at $n = 11$, $k = 0$ (**a–d**), $k = 2$ (**e–h**), $k = 5$ (**i–l**), $m = 1$, $a = 1$, $w_0 = 500$ µm in the initial plane $z = 0$ (**a**, **b**, **e**, **f**, **i**, **j**) and in the far field $z = 2z_0$ (**c**, **d**, **g**, **h**, **k**, **l**). The red circles in phase distributions (**d**, **h**, **l**) highlight the optical vortices. Scale mark in each figure means 1 mm

vortices becomes integer ($TC = nm$) and is conserved on further propagation. Earlier, in [43], it was shown that if a spiral phase plate with the transmittance $\exp(in\varphi)$ is fabricated for a specific wavelength λ_0 and illuminated by a laser light with another wavelength λ, then the TC n becomes fractional $n(\lambda_0/\lambda)$ and, on propagation in space, according to [6], the number of vortices should double and the TC should become integer and equal to $2n$ (if $\lambda < \lambda_0$). This is, however, only approximately, since when an optical vortex with a fractional TC in the initial plane propagates in space, its TC becomes undefined [40]. For the geometric progression of optical vortices, if the initial TC is half-integer ($TC = n/2$), then, on propagation, the TC acquires a definite integer value, which is twice higher ($TC = n$).

5.3 Common Topological Charge of a Superposition of Several Identical Off-Axis Vortex Beams

Optical vortices are characterized by their orbital angular momentum (OAM) [5, 44–46] and by the topological charge (TC) [6]. For rotationally symmetric light fields, the TC value coincides with the OAM value normalized by the field intensity. Structured light of more general form can be characterized by these quantities separately, as well as by other propagation-invariant quantities such as the angular harmonics spectrum (OAM-spectrum) [47, 48], partial TC [49], OAM moments of higher orders [50].

A lot of works exists on measuring the TC of the conventional rotationally symmetric optical and acoustic vortices [51–54], but for arbitrary-shape beams it should be measured according to the definition given by Berry [6], like, for instance, in [55] by using the Shack-Hartmann wavefront sensor. In contrast to the OAM and OAM-spectrum, the TC is hard to measure since it depends on the phase distribution in the beam periphery. However, its advantage is its discreteness, i.e. it changes stepwise when the distortions achieve some level. Therefore, the TC of structured light beams receives attention even nowadays.

For example, in work [56], off-axis two-vortex beams were studied and generated. Besides studying how their intensity shape changes with changing the distance between the vortices, TC of such off-axis two-vortex beams was determined by using a tilted lens. It was demonstrated that the TC depends not only on the magnitudes of the azimuthal indices of the both phase singularities, but also on their signs. It was also shown that a tilted lens allows determining the spatial distribution of the vortices.

In works [57, 58] TC of fractional-order optical vortices is investigated in the far field. This work is a continuation of works [59–61] on studying the TC of fractional vortices in the far field, but it studies also transition curves between two integer TC values, which, as turns out, depend also on the waist radius of the vortex beam.

Besides studying the TC and the OAM of various light fields and ways of their measurements, some works suggested using these quantities for optical computations [21, 62].

In our previous works, we also demonstrated how simple arithmetic operations with the TC can be implemented. For example, if the Laguerre-Gaussian beam is added with an off-axis Gaussian beam, the TC of the superposition is equal to the TC of the Laguerre-Gaussian beam, divided by 2 (integer division) [63]. In [19], we shown that choosing the phase delays between parallel Laguerre-Gaussian beams, the TC of such a superposition can be made equal to the TC of each constituent beam, but incremented or decremented by a unit. In this case, the beams should be located in the transverse plane on a circle. Several other works also studied vortex beams formed from an array of Gaussian beams with phase delays between them [64–66]. In [64, 65], a superposition of beams was considered that comes from a hexagonal matrix of fiber lasers, and propagation dynamics of such a synthesized vortex beam was compared with the Laguerre-Gaussian beam both in free space and in a turbulent medium. In [66], Gaussian beams were located on a circle, and a general expression was obtained for the complex amplitude of such a superposition. This expression was used for computing the intensity and phase distributions, necessary for observing the optical vortex generation, but the vortex order was not obtained analytically by the derived formula.

An array of single-ringed Laguerre-Gaussian, rather than an array of the Gaussian beams, with their axes located on a surface of a hyperboloid of revolution, was investigated in [67]. Such arrays of identical beams can be generated by a coherent beam combining method [25]. A general expression was obtained for the OAM and its partial cases were analyzed. It was shown in [67] that the OAM can achieve arbitrary large values limited only by the vortices density, but the TC of the whole superposition was not studied in this work. The work [68] also deals with an array of single-ringed Laguerre-Gaussian beams with their axes on the surface of a hyperboloid of revolution. For such an array, central critical point was studied (central vortex) and its TC, but the common TC of the whole transverse distribution was not considered and, in addition, only superposition of the Laguerre-Gaussian beams with only one ring was studied. The work [69] also studies arrays of identical beams, but these beams are vectorial and generated inside a laser cavity. The considered beam families are constructed of several ray-like trajectories with wave properties. Such fields possess four degrees of freedom and can be transformed into vortex beams with a large OAM value, with several singularities, or into helical star-like structures. For studying polarization singularities and topological phase of such fields, distributions of two phase parameters was considered: arctan($|E_x|/|E_y|$) and arg(E_x/E_y) with E_x and E_y being the field components, although the integral values of these quantities for the whole field, such as the OAM or TC, were not obtained analytically.

In this section, we investigate a superposition of arbitrary-shaped off-axis optical vortices, not necessarily the Laguerre-Gaussian beams. We show for such superpositions that choosing the phase delays between the beams allows changing the TC and, in addition, if the number of the constituent beams increases, the TC can be changed by a greater value, increasing it by 2, 3, or 4. Theoretical prove of changing TC is given for the far field of diffraction, but these theoretical TC values are confirmed numerically for the near field as well.

5.3.1 Topological Charge of a Superposition of Identical Parallel Vortex Beams in the Far Field of Diffraction

A superposition of identical parallel vortex beams has the following complex amplitude in the initial plane:

$$E(x, y, 0) = \sum_{n=0}^{N-1} c_n A\left\{\left[(x - a_n)^2 + (y - b_n)^2\right]^{1/2}\right\} \times \exp\left\{im \arg\left[(x - a_n) + i(y - b_n)\right]\right\}. \quad (5.44)$$

where (x, y) are the Cartesian coordinates in the initial plane, N is the number of beams in the superposition, m is the TC of each constituent vortex beam, (a_n, b_n) are the Cartesian coordinates of the nth beam's center, A is a function of the radial polar coordinate defining the transverse beam shape (Laguerre-Gaussian [70, 71], or Bessel-Gaussian [32, 72], or other), c_n are the superposition weight coefficients, i.e. $|c_n|$ is the amplitude of nth constituent beam, while arg c_n is its initial phase.

In the far field, complex amplitude can be obtained by the Fourier transform [73] of the initial amplitude:

$$E(x, y, z >> z_0) = \frac{-ik}{2\pi z} \int\limits_{-\infty}^{+\infty} \int\limits_{-\infty}^{+\infty} E(x\prime, y\prime, 0) \exp\left[-i\frac{k}{z}(xx\prime + yy\prime)\right] dx\prime dy\prime, \tag{5.45}$$

where z_0 is some distance that divides the propagation axis into near ($z << z_0$) and far ($z >> z_0$) field of diffraction, for instance, the Rayleigh length for light fields with the Gaussian envelope.

Substituting Eq. (5.44) into Eq. (5.45) and using the Fourier transform property for the shifted functions, we get

$$E(x, y, z >> z_0) = \frac{-ik}{2\pi z} \sum_{n=0}^{N-1} c_n \exp\left[-i\frac{k}{z}(a_n x + b_n y)\right] \times \int\limits_{-\infty}^{+\infty} \int\limits_{-\infty}^{+\infty} A\left(\sqrt{x'^2 + y'^2}\right) \exp\left[im \arg(x\prime + iy\prime) - i\frac{k}{z}(xx\prime + yy\prime)\right] dx\prime dy\prime. \tag{5.46}$$

Now we will use the polar coordinates both in the initial plane and in the far field:

$$\begin{aligned} x &= r \cos \varphi \\ y &= r \sin \varphi \\ x\prime &= r\prime \cos \varphi\prime \\ y\prime &= r\prime \sin \varphi\prime. \end{aligned} \tag{5.47}$$

Then, after evaluating the integral over the angular polar coordinate φ' in the initial plane, the Fourier transform (5.46) reads as

$$E(r, \varphi, z >> z_0) = (-i)^{m+1} \frac{k}{z}$$

$$\times \sum_{n=0}^{N-1} c_n \exp\left[-i\frac{k}{z}r(a_n \cos\varphi + b_n \sin\varphi)\right]$$

$$\exp(im\varphi) \int_0^\infty A(r\prime)J_m\left(\frac{k}{z}r\prime r\right)r\prime dr\prime. \quad (5.48)$$

It is seen that the far-field complex amplitude of the whole superposition equal the far-field complex amplitude of a single constituent vortex beam E_{vortex}, multiplied by the complex amplitude of a superposition of plane waves E_{plane}:

$$E(r, \varphi, z >> z_0) = E_{\text{vortex}}(r, \varphi, z >> z_0) E_{\text{plane}}(r, \varphi, z >> z_0), \quad (5.49)$$

Where

$$E_{\text{vortex}}(r, \varphi, z >> z_0) = (-i)^{m+1}\frac{k}{z}\exp(im\varphi) \int_0^\infty A(r\prime)J_m\left(\frac{k}{z}r\prime r\right)r\prime dr\prime, \quad (5.50)$$

$$E_{\text{plane}}(r, \varphi, z >> z_0) = \sum_{n=0}^{N-1} c_n \exp\left[-i\frac{k}{z}r(a_n \cos\varphi + b_n \sin\varphi)\right]. \quad (5.51)$$

Then the topological charge of the superposition is equal to the sum of the topological charge of one constituent beam m and of the topological charge of a superposition of plane waves (5.51) TC_{plane}:

$$TC = m + TC_{\text{plane}} \quad (5.52)$$

Such a separation of the common TC into the TC of a single constituent beam and the TC of a certain superposition of plane waves is possible due to considering the light field in the far field of diffraction. Before the far field, the light beam propagates in the Fresnel diffraction zone. For this zone, we failed to derive the common TC of the superposition (5.44) analytically. However, the work [49] has proved the TC conservation upon propagation, even if the field intensity changes. Therefore, we will derive the TC analytically for the far field by using Eq. (5.52), while its validity for the Fresnel diffraction zone will be verified by numerical computations.

We can obtain the additive topological charge TC_{plane} in a simple partial case when $c_n = \exp(i2\pi pn/N)$, $a_n = r_0\cos(2\pi n/N)$, $b_n = r_0\sin(2\pi n/N)$. In this case, the vortex beams are located uniformly on a circle and their phases grow along this circle linearly (Fig. 5.8).

In the far field, complex amplitude of such a beam is given by

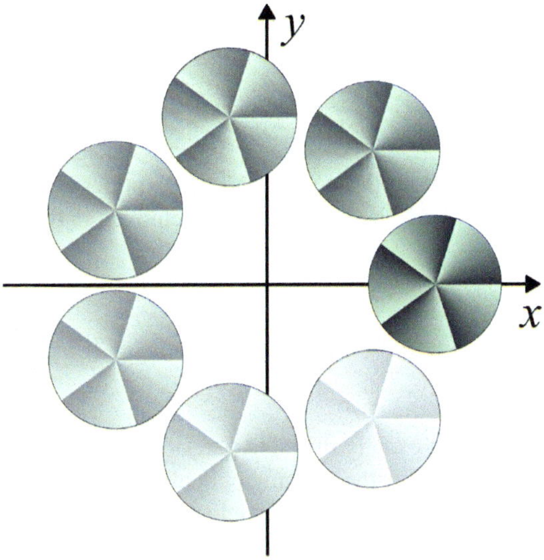

Fig. 5.8 Superposition of several optical vortices on a circle with initial phases (bright color means 0, pale color means 2π) linearly increasing along this circle

$$E_{\text{plane}}(r, \varphi, z >> z_0) = \sum_{n=0}^{N-1} \exp\left[i2\pi\frac{pn}{N} - i\frac{k}{z}r_0 r\cos\left(\varphi - \frac{2\pi n}{N}\right)\right]. \tag{5.53}$$

It is not difficult to prove the following symmetry property for a superposition of plane waves (5.53):

$$E_{\text{plane}}(r, -\varphi \pm \pi, z >> z_0) = E^*_{\text{plane}}(r, \varphi, z >> z_0). \tag{5.54}$$

Here we prove the symmetry property (5.54). For this purpose, we consider the light field on a ray with the angular polar coordinate $-\varphi \pm \pi$:

$$E_{\text{plane}}(r, -\varphi \pm \pi, z >> z_0) = \sum_{n=0}^{N-1} \exp\left[i2\pi\frac{pn}{N} - i\frac{k}{z}r_0 r\cos\left(-\varphi \pm \pi - \frac{2\pi n}{N}\right)\right]. \tag{5.55}$$

The cosine is the even function. In addition, we use the reduction formulas and change the summation range from $n = 0, \ldots, N - 1$ to $n = 1, \ldots, N$. This can be done since the terms in the sum at $n = 0$ and at $n = N$ are the same. Thus, we get

$$E_{\text{plane}}(r, -\varphi \pm \pi, z >> z_0) = \sum_{n=1}^{N} \exp\left[i2\pi\frac{pn}{N} + i\frac{k}{z}r_0 r\cos\left(\varphi + \frac{2\pi n}{N}\right)\right]. \tag{5.56}$$

Replacing the summation index from n to $N - n$ yields:

$$E_{\text{plane}}(r, -\varphi \pm \pi, z >> z_0) = \sum_{n=0}^{N-1} \exp\left[i2\pi \frac{p(N-n)}{N} + i\frac{k}{z} r_0 r \cos\left(\varphi + \frac{2\pi (N-n)}{N}\right)\right]. \tag{5.57}$$

Since both the cosine and the exponential in Eq. (5.57) are periodical with the period of 2π, the term N can be removed in the numerators and thus

$$E_{\text{plane}}(r, -\varphi \pm \pi, z >> z_0) = \sum_{n=0}^{N-1} \exp\left[-i2\pi \frac{pn}{N} + i\frac{k}{z} r_0 r \cos\left(\varphi - \frac{2\pi n}{N}\right)\right] \tag{5.58}$$

i.e. we get exactly the right part of Eq. (5.54): $E^*_{\text{plane}}(r, \varphi, z >> z_0)$.

This means that for any optical vortex, another symmetrical optical vortex exists and it has the opposite topological charge. Thus, all off-axis vortices in the field E_{plane} compensate each other and do not affect the common topological charge of the whole superposition of plane waves. Only the central vortex on the optical axis is not compensated. To determine its topological charge, the superposition of plane waves can be expanded into a Taylor series:

$$E_{\text{plane}}(r, \varphi, z >> z_0) = \sum_{n=0}^{N-1} \exp\left(i2\pi \frac{pn}{N}\right) \sum_{s=0}^{\infty} \frac{1}{s!} \left(-i\frac{k}{z} r_0 r\right)^s \cos^s\left(\varphi - \frac{2\pi n}{N}\right). \tag{5.59}$$

Representing the cosines as sums of exponentials, applying then the binomial theorem to them, and using the following identity

$$\sum_{n=0}^{N-1} \exp\left(i2\pi \frac{pn}{N}\right) = N\delta_p, \tag{5.60}$$

where $\delta_p = 1$ for $p = 0, N, 2N, \ldots$ and $\delta_p = 0$ for other values p, Eq. (5.59) is transformed to the following form:

$$E_{\text{plane}}(r, \varphi, z >> z_0) = N \sum_{s=0}^{\infty} \left(-i\frac{k}{2z} r_0 r\right)^s \sum_{t=0}^{s} \frac{\delta_{p-2t+s}}{t!(s-t)!} \exp[i(2t-s)\varphi]. \tag{5.61}$$

It is seen that nonzero terms in this sum are those that describe optical vortices with the topological charge of p. Due to ciclicity of the values δ_p, other terms in the sum (14) can be also nonzero. These terms describe optical vortices with the topological charges $p \pm N$, $p \pm 2N$, …, but their contribution decays due to the coefficients $1/[t!(s-t)!]$. Thus, the topological charge of the whole superposition should be equal to

$$TC = m + p. \tag{5.62}$$

For this reason, we call the parameter p a topological charge increment. We note that if the field is composed of optical vortices, located on several circles, similarly to several hexagonal rings in [64], then it is described by a sum of several terms, each of the form (5.53). Each of these terms has the symmetry (5.54) and has the central optical vortex with the TC given by Eq. (5.62). Thus, as has been shown in [64], the number of rings with the vortices affects the shape of the generated vortex beam, but the common TC of the whole superposition remains unchanged with changing the number of rings and is equal to (5.62).

Further, we verify the formula (5.62) by the numerical computation for several different light fields.

5.3.2 Numerical Simulation

Shown in Fig. 5.9 are the intensity and phase distributions of several superpositions of eight 7th-order LG beams with a different topological charge increment from −2 till +3.

The topological charge was computed along a large-radius circle ($R = 30$ mm) so as to bypass the whole area of significant intensity. Computation yields the following TC values: $4.9455 \approx 5$ (Fig. 5.9a, d), $5.9333 \approx 6$ (Fig. 5.9b, e), $6.9218 \approx 7$ (Fig. 5.9c, f), $7.9103 \approx 8$ (Fig. 5.9 g, j), $8.8988 \approx 9$ (Fig. 5.9 h, k), $9.8870 \approx 10$ (Fig. 5.9i, l). Thus, all these cases confirm that the topological charge equals $TC = m + p$.

Computation of the topological charge along the same circle, as for Fig. 5.9, yields the following TC values: $4.9171 \approx 5$ (Fig. 5.10a, d), $5.8991 \approx 6$ (Fig. 5.10b, e), $6.8818 \approx 7$ (Fig. 5.10c, f), $7.8645 \approx 8$ (Fig. 5.10 g, j), $8.8472 \approx 9$ (Fig. 5.10 h, k), $9.8297 \approx 10$ (Fig. 5.10i, l). Thus, in all these cases, the topological charge value $TC = m + p$ is confirmed for the Bessel-Gaussian beams as well. Intensity and phase distributions of several superpositions of eight Bessel-Gaussian beams are illustrated in Fig. 5.10. As in Fig. 5.9, these beams also have 7th order and the topological charge increments from −2 till +3.

Now we determine whether the distance between the beams affects the common topological charge. For this purpose, we increase the distance from the beams centers to the optical axis a_0.

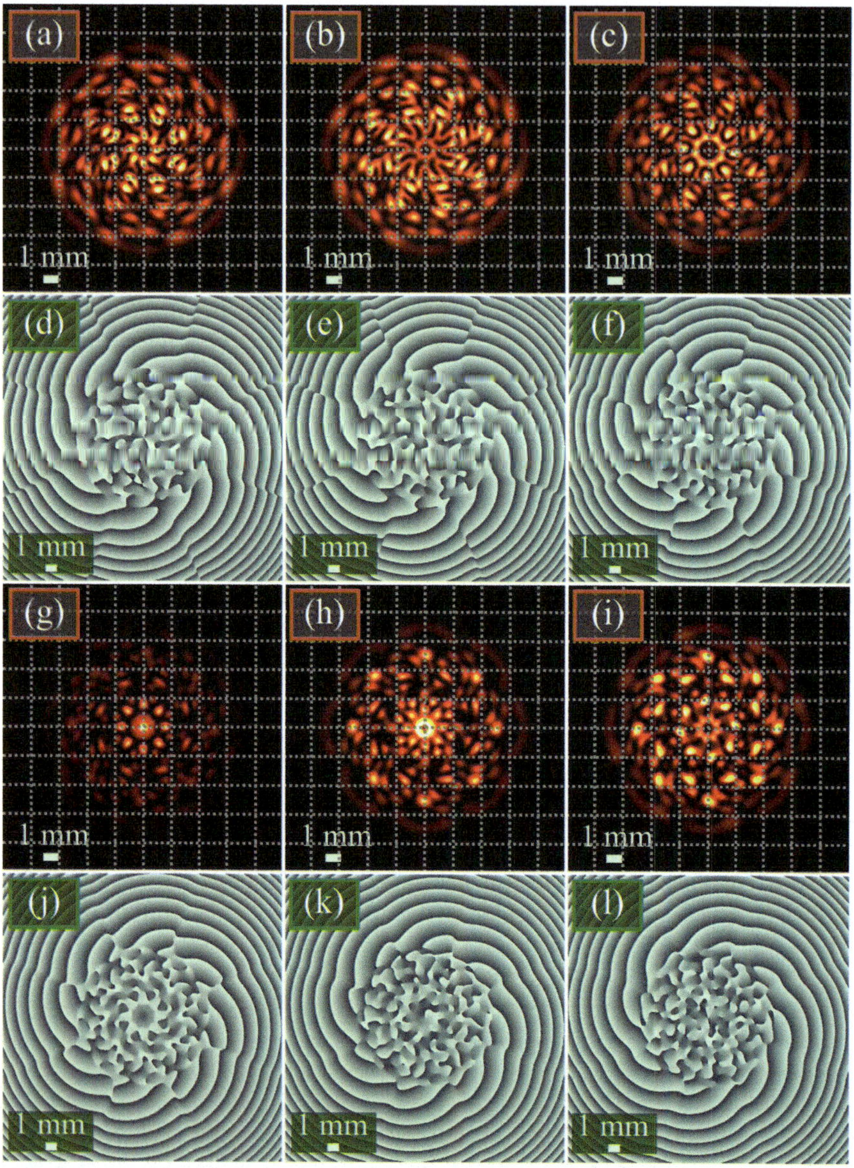

Fig. 5.9 Intensity (**a**–**c**, **g**–**i**) and phase (**d**–**f**, **j**–**l**) distributions of several superpositions of the LG beams for the following parameters: wavelength $\lambda = 532$ nm, waist radius of the Gaussian beam $w_0 = 1$ mm, radial and azimuthal indices of each constituent LG beam $q = 2$ and $m = 7$, number of beams $M = 8$, distance from the beams centers to the optical axis $a_0 = 3w_0 = 3$ mm, propagation distance $z = z_0/2$, topological charge increment $p = -2$ (**a**, **d**), -1 (**b**, **e**), 0 (**c**, **f**), +1 (**g**, **j**), +2 (**h**, **k**), +3 (**i**, **l**)

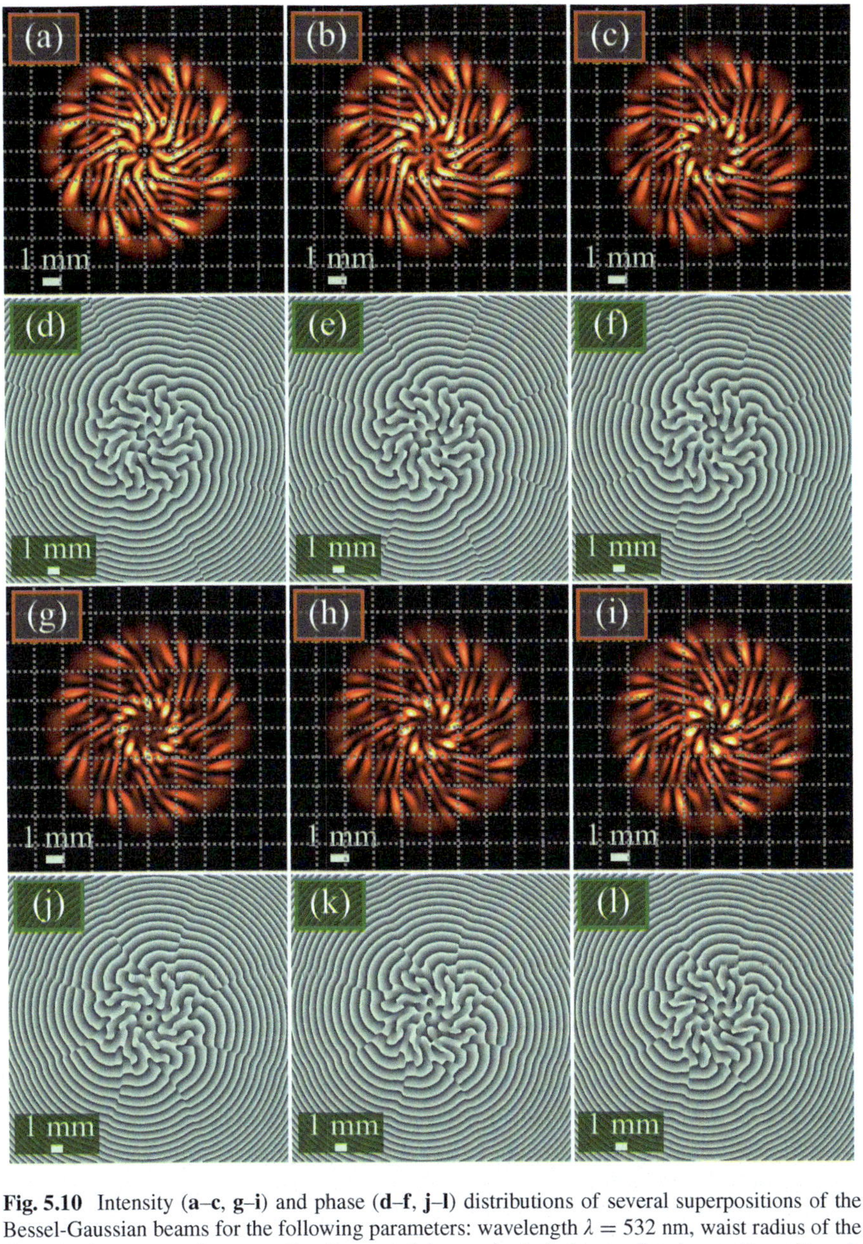

Fig. 5.10 Intensity (**a–c**, **g–i**) and phase (**d–f**, **j–l**) distributions of several superpositions of the Bessel-Gaussian beams for the following parameters: wavelength $\lambda = 532$ nm, waist radius of the Gaussian beam $w_0 = 1$ mm, scaling factor and the topological charge of each constituent Bessel-Gaussian beam $\alpha = k/2000$ and $m = 7$, number of beams $M = 8$, distance from the beams centers to the optical axis $a_0 = 3w_0 = 3$ mm, propagation distance $z = z_0/2$, topological charge increment $p = -2$ (**a, d**), -1 (**b, e**), 0 (**c, f**), +1 (**g, j**), +2 (**h, k**), +3 (**i, l**)

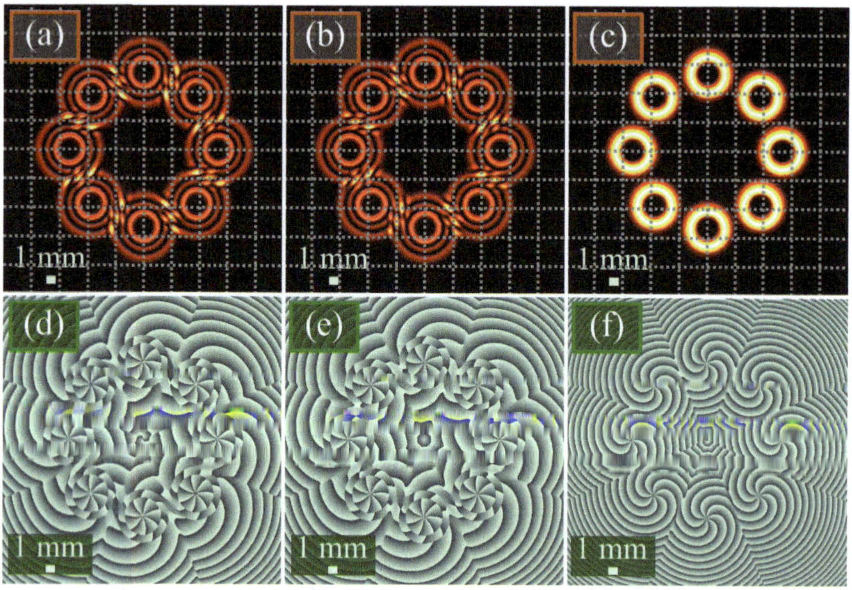

Fig. 5.11 Intensity (**a–c**) and phase (**d–f**) distributions of several superpositions of the LG beams (**a**, **b**, **d**, **e**) and the Bessel-Gaussian (**c**, **f**) beams for the following parameters: wavelength $\lambda =$ 532 nm, waist radius of the Gaussian beam $w_0 = 1$ mm, radial index of the LG beams $q = 2$, scaling factor of the Bessel-Gaussian beams $\alpha = k/2000$, topological charge of each constituent beam in the superposition $m = 7$, number of beams $M = 8$, distance from the beams centers to the optical axis $a_0 = 8w_0 = 8$ mm, propagation distance $z = z_0/2$, topological charge increment $p = 3$ (**a**, **d**), 1 (**b**, **e**), 2 (**c**, **f**)

Figure 5.11 depicts the intensity and phase distributions of several superpositions of eight LG and Bessel-Gaussian beams when the distance from the beams centers to the optical axis equals $a_0 = 8w_0$ (in Figs. 5.9 and 5.10, a0 $= 3w_0$).

Computation of the topological charge yields the following values: 9.9256 $\approx$ 10 (Fig. 5.11a, d), 7.9425 $\approx$ 8 (Fig. 5.11b, e), 8.8911 $\approx$ 9 (Fig. 5.11c, f). Thus, a superposition of beams, more remote from the optical axis, has the topological charge, also equal to $TC = m + p$.

Now we compute the TC for a light field with two rings with the optical vortices. Shown in Fig. 5.12 are the intensity and phase distributions of a superposition of sixteen LG beams, residing on two rings (eight beams on each ring) with the radii $a_{01} = 5w_0$ and $a_{02} = 10w_0$.

In the near field (Fig. 5.12a, d), LG beams are clearly seen on the outer ring, as well as clearly seen the central dark areas of the LG beams on the inner ring. When passing from the near field to the far field, at the Rayleigh distance (Fig. 5.12b, e), LG beams on the outer ring can be distinctly seen, whereas the LG beams on the inner ring are undistinguishable due to interference. In the far field (Fig. 5.12c, f), due to interference, the Laguerre-Gaussian beams are undistinguishable both in the outer and on the inner ring.

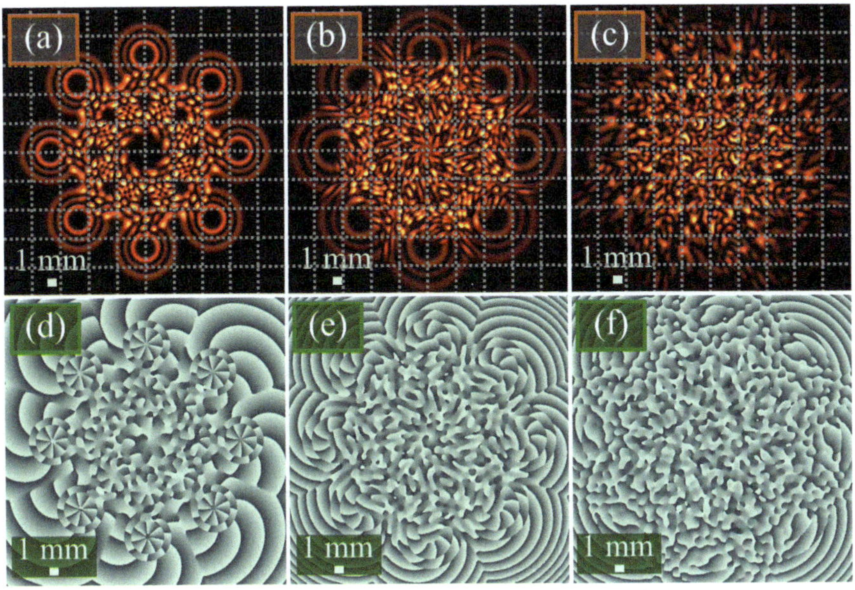

Fig. 5.12 Intensity (**a**–**c**) and phase (**d**–**f**) distributions of a superposition of the LG beams, residing on two rings, for the following parameters: wavelength $\lambda = 532$ nm, waist radius of the Gaussian beam $w_0 = 1$ mm, radial index of the Laguerre-Gaussian beams $q = 2$, topological charge of each constituent beam in the superposition is $m = 7$, number of beams $M = 2 \times 8 = 16$, distance from the beams centers to the optical axis $a_{01} = 5w_0 = 5$ mm (inner ring) and $a_{02} = 10w_0 = 10$ mm (outer ring), topological charge increment $p = 2$, propagation distance $z = z_0/5$ (**a**, **d**) (near field), $z = z_0$ (**b**, **e**), $z = 2z_0$ (**c**, **f**) (far field)

Despite the radical change of the diffraction pattern, the computed topological charge values in all three planes are nearly the same: $8.9512 \approx 9$ (Fig. 5.12a, d), $8.9172 \approx 9$ (Fig. 5.12b, e), $8.9473 \approx 9$ (Fig. 5.12c, f). Thus, for a superposition of light beams, located on two rings, we get the topological charge which is also equal to $TC = m + p$.

We have studied the common topological charge of a superposition of several parallel optical vortices with an arbitrary radial amplitude distribution. For the far field of diffraction, we have proved theoretically that if the vortices are located uniformly on a circle then the topological charge of the superposition can differ from the topological charge of each constituent vortex beam by a given value due to the phase difference between the beams. The topological charge of an arbitrary light field should not change upon propagation since it characterizes the topology of the field phase distribution. Changing this topology requires some nonlinear transforms, and thus we suppose that, in homogeneous medium, the above topological charge of the superposition should be conserved also in the near field. The numerical simulation, conducted for multiple-ringed LG beams and for the Bessel-Gaussian beams, has confirmed that the formula $TC = m + p$ is valid even in the near field, at a distance from the initial plane, two times shorter than the Rayleigh length. A limitation of the considered model is the requirement that the beams should be parallel, i.e. the common TC of the beams array cannot be determined if they have some tilts to the optical axis. For instance, the model does not allow obtaining the TC of a structurally stable superposition similar to that studied in [68]. On the contrary, universality of the considered model is independence of the shape of the constituent vortex beams [74], since it is not supposed that they are the simplest single-ringed LG beams. The obtained feature of superpositions of parallel identical optical vortices can be used for optical data transmission if the incoming optical signals are identified by their topological charge, as well as for controlling the topological charge of a combined beam by inducing phase delays between the constituent beams.

References

1. Y. Shen, X. Wang, Z. Xie, C. Min, X. Fu, Q. Liu, M. Gong, X. Yuan, Optical vortices 30 years on: OAM manipulation from topological charge to multiple singularities. Light Sci. Appl. **8**, 90 (2019)
2. X.J. Lin, Q.C. Feng, Y. Zhu, S.H. Ji, B. Xiao, H.Y. Xu, W.S. Li, Z.P. Cai, Diode-pumped wavelength-switchable visible Pr^{3+}:YLF laser and vortex laser around 670 nm. Opto-Electron. Adv. **4**, 210006 (2021)
3. D.S. Zhang, X.Z. Li, Y. Fu, Q.H. Yao, Z.G. Li, K. Sugioka, Liquid vortexes and flows induced by femtosecond laser ablation in liquid governing formation of circular and crisscross LIPSS. Opto-Electron. Adv. **5**, 210066 (2022)
4. Y. Zhang, X. Liu, H. Lin, D. Wang, E. Cao, S. Liu, Z. Nie, B. Jia, Ultrafast multi-target control of tightly focused light fields. Opto-Electr. Adv. **5**, 210026 (2022)
5. L. Allen, M. Beijersbergen, R. Spreeuw, J. Woerdman, Orbital angular momentum of light and the transformation of Laguerre-Gaussian laser modes. Phys. Rev. A **45**, 8185–8189 (1992)
6. M.V. Berry, Optical vortices evolving from helicoidal integer and fractional phase steps. J. Opt. A Pure Appl. Opt. **6**, 259–268 (2004)
7. G. Molina-Terriza, J. Recolons, L. Torner, The curious arithmetic of optical vortices. Opt. Lett. **25**, 1135–1137 (2000)
8. V. Pyragaite, V. Smilgevicius, A. Stabinis, V. Maslinska, The interference of higher order Laguerre-Gaussian beams. Lithuan. J. Phys. **47**, 21–26 (2007)
9. V. Pyragaite, V. Stabinis, Free-space propagation of overlapping light vortex beams. Opt. Commun.Commun. **213**, 187–191 (2002)
10. D. Lopez-Mago, B. Perez-Garcia, A. Yepiz, R.I. Hernandez-Aranda, J.C. Gutiérrez-Vega, Dynamics of polarization singularities in composite optical vortices. J. Opt. **15**, 044028 (2013)
11. D.N. Naik, N.K. Viswanathan, Generation of singular optical beams from fundamental Gaussian beam using Sagnac interferometer. J. Opt. **18**, 095601 (2016)
12. K. Cheng, B. Lü, Composite coherence vortices in coherent and incoherent superpositions of two off-axis partially coherent vortex beams. J. Mod. Opt. **55**, 2751–2764 (2008)
13. A.P. Sukhorukov, A.A. Kalinovich, G. Molina-Terriza, L. Torner, Superposition of noncoaxial vortices in parametric wave mixing. Phys. Rev. E **66**, 036608 (2002)
14. Y. Zhang, Z. Wu, K. Yang, P. Li, F. Wen, Y. Gu, Splitting, generation, and annihilation of phase singularities in non-coaxial interference of Bessel-Gaussian beams. Phys. Scr. **96**, 125105 (2021)
15. V.V. Kotlyar, A.A. Kovalev, P. Amiri, P. Soltani, S. Rasouly, Topological charge of two parallel Laguerre-Gaussian beams. Opt. Express **29**, 42962–42977 (2021)
16. A.A. Kovalev, V.V. Kotlyar, Orbital angular momentum of superposition of identical shifted vortex beams. J. Opt. Soc. Am. A **32**, 1805–1810 (2015)

17. V.V. Kotlyar, A.A. Kovalev, *Topological Charge of Optical Vortices* (Novaya Technika, Samara, 2021)
18. V.V. Kotlyar, A.A. Kovalev, A.V. Volyar, Topological charge of a linear combination of optical vortices: topological competition. Opt. Express **28**, 8266–8281 (2020)
19. V.V. Kotlyar, A.A. Kovalev, E.S. Kozlova, A.A. Savelyeva, Tailoring the topological charge of a superposition of identical parallel Laguerre-Gaussian beams. Micromachines **13**, 2227 (2022)
20. Q. Zhang, J. Ni, C.W. Qiu, Vortex 4.0 on chip. Light Sci. Appl. **9**, 103 (2020)
21. A.M. Akulshin, I. Novikova, E.E. Mikhailov, S.A. Suslov, R.J. McLean, Arithmetic with optical topological charges in stepwise-excited Rb vapor. Opt. Lett.Lett. **41**, 1146–1149 (2016)
22. I.J. Lee, S. Kim, Directional coupler design for orbital angular momentum mode-based photonic integrated circuits. Opt. Express **28**, 30085 (2020)
23. I.J. Lee, S. Kim, On-chip guiding of higher-order orbital angular momentum modes. Photonics **6**, 72 (2019)
24. J. Chen, C. Wan, A. Chong, Q. Zhan, Experimental demonstration of cylindrical vector spatiotemporal optical vortex. Nanophotonics **10**, 4489–4495 (2021)
25. J. Long, T. Hou, Q. Chang, T. Yu, R. Su, P. Ma, Y. Ma, P. Zhou, L. Si, Generation of optical vortex lattices by a coherent beam combining system. Opt. Lett. **46**, 3665–3668 (2021)
26. D.A. Ikonnikov, S.A. Myslivets, V.G. Arkhipkin, A.M. Vyunishev, 3D optical vortex lattices. Ann. Phys. **533**, 2100114 (2021)
27. Y. Yang, Y. Ren, M. Chen, Y. Arita, C. Rosales-Guzmán, Optical trapping with structured light: a review. Adv. Photon. **3**, 034001 (2021)
28. C. Li, S.A. Maier, H. Ren, Optical vortices in nanophotonics. Chin. Opt. **14**, 792–811 (2021)
29. A. Pryamikov, L. Hadzievski, M. Fedoruk, S. Turitsyn, A. Aceves, Optical vortices in waveguides with discrete and continuous rotational symmetry. J. Eur. Opt. Soc. Rapid Publ. **17**, 23 (2021)
30. A.V. Volyar, E.G. Abramochkin, Y.E. Akimova, M.V. Bretsko, Reconstruction of stable states of spiral vortex beams. Comput. Opt.. Opt. **46**, 5–15 (2022)
31. J. Durnin, J.J. Miceli Jr., J.H. Eberly, Diffraction-free beams. Phys. Rev. Lett. **58**, 1499–1501 (1987)
32. F. Gori, G. Guattari, C. Padovani, Bessel-Gauss beams. Opt. Commun.Commun. **64**, 491–495 (1987)
33. V.V. Kotlyar, A.A. Almazov, S.N. Khonina, V.A. Soifer, H. Elfstrom, J. Turunen, Generation of phase singularity through diffracting a plane or Gaussian beam by a spiral phase plate. J. Opt. Soc. Am. A **22**, 849–861 (2005)
34. V.V. Kotlyar, R.V. Skidanov, S.N. Khonina, V.A. Soifer, Hypergeometric modes. Opt. Lett. **32**, 742–744 (2007)
35. E. Karimi, G. Zito, B. Piccirillo, L. Marrucci, E. Santamato, Hypergeometric-Gaussian modes. Opt. Lett. **32**, 3053–3055 (2007)
36. M.A. Bandres, J.C. Gutiérrez-Vega, Circular beams. Opt. Lett. **33**, 177–179 (2008)
37. M.A. Bandres, J.C. Gutiérrez-Vega, Elliptical beams. Opt. Express **16**, 21087–21092 (2008)
38. V.V. Kotlyar, A.A. Kovalev, V.A. Soifer, Asymmetric Bessel modes. Opt. Lett. **39**, 2395–2398 (2014)
39. E.G. Abramochkin, V.G. Volostnikov, Generalized Gaussian beams. J. Opt. A Pure Appl. Opt. **6**, S157–S161 (2004)
40. A.G. Nalimov, V.V. Kotlyar, Topological charge of optical vortices in the far field with an initial fractional charge: optical "dipoles." Comput. Opt.. Opt. **46**, 189–195 (2022)
41. V.V. Kotlyar, A.A. Kovalev, Optical vortex beams with a symmetric and almost symmetric OAM spectrum. J. Opt. Soc. Am. A **38**, 1276–1283 (2021)
42. V. Kotlyar, A. Kovalev, E. Kozlova, A. Savelyeva, S. Stafeev, Geometric progression of optical vortices. Photonics **9**, 407 (2022)
43. G. Gbur, Fractional vortex Hilbert's hotel. Optica **3**, 222–225 (2016)
44. A.V. Volyar, E.G. Abramochkin, M.V. Bretsko, S.I. Khalilov, Y.E. Akimova, Control of giant orbital angular momentum bursts of structured Laguerre-Gaussian beams in a medium with general astigmatism. Comput. Opt.. Opt. **48**, 35–46 (2024)

45. M.V. Bretsko, Y.E. Akimova, A.V. Volyar, S.I. Khalilov, Y.A. Egorov, A.O. Ivakhnenko, Orbital angular momentum of structured LG beams after astigmatic transformation. Opt. Mem. Neural Netw. **32**, S75–S83 (2023)
46. C.N. Alexeyev, S.S. Alieva, E.V. Barshak, B.P. Lapin, M.A. Yavorsky, Structured light transformations and orbital angular momentum control in a three-coil optical snake. Comput. Opt.. Opt. **46**, 701–712 (2022)
47. A. Volyar, M. Bretsko, Y. Akimova, Y. Egorov, Measurement of the vortex and orbital angular momentum spectra with a single cylindrical lens. Appl. Opt. **58**, 5748–5755 (2019)
48. A. Volyar, M. Bretsko, Y. Akimova, Y. Egorov, Measurement of the vortex spectrum in a vortex-beam array without cuts and gluing of the wavefront. Opt. Lett. **43**, 5635–5638 (2018)
49. A.A. Kovalev, V.V. Kotlyar, A.G. Nalimov, Topological charge and asymptotic phase invariants of vortex laser beams. Photonics **8**, 445 (2021)
50. D.H. Le, A. Pal, A. Qadeer, M. Kleinert, J. Kleinert, S. Goel, K. Khare, M. Bhattacharya, Conservation of extremal ellipticity for coherent single mode Gaussian beams propagating in rotationally invariant media. Opt. Commun.Commun. **503**, 127465 (2022)
51. M.E. Anderson, H. Bigman, L.E. de Araujo, J.L. Chaloupka, Measuring the topological charge of ultrabroadband, optical vortex beams with a triangular aperture. J. Opt. Soc. Am. B **29**, 1968–1976 (2012)
52. P. Vaity, J. Banerji, R.P. Singh, Measuring the topological charge of an optical vortex by using a tilted convex lens. Phys. Lett. A **377**, 1154–1156 (2013)
53. J. Zhu, P. Zhang, Q. Li, F. Wang, C. Wang, Y. Zhou, J. Wang, H. Gao, L.C. Kwek, F. Li, Measuring the topological charge of orbital angular momentum beams by utilizing weak measurement principle. Sci. Rep. **9**, 7993 (2019)
54. H. Ge, Z.W. Long, X.Y. Xu, J.C. Hua, Y. Liu, B.Y. Xie, J.H. Jiang, M.H. Lu, Y.F. Chen, Direct measurement of acoustic spectral density and fractional topological charge. Phys. Rev. Appl. **19**, 034073 (2023)
55. D. Wang, H. Huang, H. Toyoda, H. Liu, Topological charge detection using generalized contour-sum method from distorted donut-shaped optical vortex beams: experimental comparison of closed path determination methods. Appl. Sci. **9**, 3956 (2019)
56. M. Guo, W. Le, C. Wang, G. Rui, Z. Zhu, J. He, B. Gu, Generation, topological charge, and orbital angular momentum of off-axis double vortex beams. Photonics **10**, 368 (2023)
57. E. Peters, G. Funes, L. Martínez-León, E. Tajahuerce, Dynamics of fractional vortex beams at Fraunhofer diffraction zone. Photonics **9**, 479 (2022)
58. E. Peters, G. Funes, L. Martínez-León, E. Tajahuerce, Analysis of practical fractional vortex beams at far field. Opt. Laser Technol. **156**, 108480 (2022)
59. A.J. Jesus-Silva, E.J.S. Fonseca, J.M. Hickmann, Study of the birth of a vortex at Fraunhofer zone. Opt. Lett. **37**, 4552–4554 (2012)
60. J. Wen, L.C. Wang, X. Yang, J. Zhang, S.Y. Zhu, Vortex strength and beam propagation factor of fractional vortex beams. Opt. Express **27**, 5893–5904 (2019)
61. V. Kotlyar, A. Kovalev, A. Nalimov, A. Porfirev, Evolution of a vortex with an initial fractional topological charge. Phys. Rev. A **102**, 023516 (2020)
62. F. Meng, X.G. Wei, Y.J. Qu, Y. Chen, X.J. Zhang, Z.H. Kang, L. Wang, H.H. Wang, J.Y. Gao, Arithmetic operation of orbital angular momentum of light via slow-light four-wave mixing. J. Lumin.Lumin. **242**, 118551 (2022)
63. A.A. Kovalev, V.V. Kotlyar, E.S. Kozlova, M.A. Butt, Dividing the topological charge of a Laguerre-Gaussian beam by 2 using an off-axis Gaussian beam. Micromachines **13**, 1709 (2022)
64. V.P. Aksenov, V.V. Dudorov, V.V. Kolosov, Properties of vortex beams formed by an array of fibre lasers and their propagation in a turbulent atmosphere. Quant. Electr. **46**, 726 (2016)
65. V.P. Aksenov, V.V. Dudorov, V.V. Kolosov, M.E. Levitsky, Synthesized vortex beams in the turbulent atmosphere. Front. Phys. **8**, 143 (2020)
66. L.G. Wang, L.Q. Wang, S.Y. Zhu, Formation of optical vortices using coherent laser beam arrays. Opt. Commun.Commun. **282**, 1088–1094 (2009)

67. Y. Izdebskaya, T. Fadeyeva, V. Shvedov, A. Volyar, Vortex-bearing array of singular beams with very high orbital angular momentum. Opt. Lett. **31**, 2523–2525 (2006)
68. Y. Izdebskaya, V. Shvedov, A. Volyar, Symmetric array of off-axis singular beams: spiral beams and their critical points. J. Opt. Soc. Am. A **25**, 171–181 (2008)
69. Y. Shen, X. Yang, D. Naidoo, X. Fu, A. Forbes, Structured ray-wave vector vortex beams in multiple degrees of freedom from a laser. Optica **7**, 820–831 (2020)
70. H. Kogelnik, T. Li, Laser beams and resonators. Appl. Opt. **5**, 1550–1567 (1966)
71. M. Kirilenko, D. Gorelykh, Simulation of the propagation of rotating Laguerre-Gaussian beams in an imaging system with an obstacle. Opt. Mem. Neural Netw. **32**, S38–S45 (2023)
72. Y.P. Arul-Teen, T. Nathiyaa, K.B. Rajesh, S. Karthick, Bessel Gaussian beam propagation through turbulence in free space optical communication. Opt. Mem. Neural Netw. **27**, 81–88 (2018)
73. J.W. Goodman, *Introduction to Fourier Optics* (McGraw-Hill, New York, 1996)
74. A.A. Kovalev, V.V. Kotlyar, A.G. Nalimov, Common topological charge of a superposition of several identical off-axis vortex beams with an arbitrary circularly symmetric transverse shape. Opt. Memory Neural Netw. **33**, S285–S294 (2024)

Chapter 6
Orbital Angular Momentum of Helical Laser Beams

6.1 Orbital Angular Momentum of Paraxial Propagation-Invariant Laser Beams

Last years, structured light attracts much attention. There are several review papers that describe generation and detection of such light [1–3]. Such light fields can be generated both by using lasers [4, 5] and by adopting out-of-resonator optical components (for instance, an astigmatic converter [6]). Structured light is a light field that has some structure, i.e. either transverse spatial structure (for example, Hermite-Gaussian (HG) or Laguerre-Gaussian beams [7]), or longitudinal (for example, Airy beams [8]), polarizational (for example, radially or azimuthally polarized beams [9]), temporal (for example, space–time optical vortices [10, 11]) structure. Propagation-invariant (or structurally stable) beams is a subclass of structured beams and they can be represented in different ways, for example, by embedding a set of shifted optical vortices of the same sign into a Gaussian beam [12], or by using a finite superposition of the HG beams [13], or using a series [14], or by adopting a 4×4 propagation matrix [15]. Propagation-invariant laser beams are paraxial beams, whose transverse intensity structure is conserved on propagation, changing only in scale. Such beam can rotate on propagation in free space. Interestingly, in [16], the authors consider a finite superposition of the HG beams with complex coefficients to analyze the perturbation of the Laguerre-Gaussian mode. In fact, in [16], a special type of perturbation of the Laguerre-Gaussian beam is studied, namely, such that the perturbed beams still remain propagation-invariant.

In this section, we investigate propagation-invariant laser beams composed by a finite number of the HG beams. We obtain a general expression for the orbital angular momentum (OAM) of such beams and give several examples of choosing the weight coefficients of the superposition of the HG beams that yield the different OAM values. We show that the maximal OAM normalized to the beam power is equal to the maximal power of the Hermite polynomial in this sum.

V. V. Kotlyar et al., *Helical Laser Beams*, https://doi.org/10.1007/978-3-031-92217-6_6

6.1.1 Derivation of an Expression for the Orbital Angular Momentum of Propagation-Invariant Beams

Under the propagation-invariant (they are also called structured Gaussian) beams, we mean here paraxial beams composed of a finite superposition of the HG beams with a constant sum of indices equal to N:

$$
\begin{aligned}
SHG_N(x, y) = {} & \exp\left(-\frac{x^2 + y^2}{w^2}\right) \\
& \times \sum_{p=0}^{N} C_p H_{N-p}\left(\frac{\sqrt{2}x}{w}\right) H_p\left(\frac{\sqrt{2}y}{w}\right),
\end{aligned} \tag{6.1}
$$

where w is the waist radius of the Gaussian beam, C_p are the complex valued coefficients, $H_p(x)$ are the Hermite polynomials obeying the following recurrent identities:

$$
\begin{aligned}
& H_n(x) = 2xH_{n-1}(x) - 2(n-1)H_{n-2}(x), \\
& H_0(x) = 1, \quad H_1(x) = 2x.
\end{aligned} \tag{6.2}
$$

The HG beams are propagation-invariant i.e. their transverse shape is conserved on propagation and changes only in scale. The Fresnel transform conserves the HG beam:

$$
\begin{aligned}
HG_{n,m}(\xi, \eta, z) = {} & \left(-\frac{ik}{2\pi z}\right) \int_{-\infty}^{\infty}\int_{-\infty}^{\infty} \exp\left(-\frac{x^2 + y^2}{w^2}\right) \\
& \times H_n\left(\frac{\sqrt{2}x}{w}\right) H_m\left(\frac{\sqrt{2}y}{w}\right) \times \\
& \times \exp\left[\frac{ik}{2z}\left(x^2 + y^2 + \xi^2 + \eta^2 - 2x\xi - 2y\eta\right)\right] dxdy \\
= {} & \frac{1}{|q(z)|} \exp\left[\frac{iz}{z_0 w^2 |q(z)|^2}\left(\xi^2 + \eta^2\right)\right] \times \\
& \times \exp\left[-i(n + m + 1)\arctan\left(\frac{z}{z_0}\right)\right] \\
& \times \exp\left(-\frac{\xi^2 + \eta^2}{w^2 |q(z)|^2}\right) H_n\left(\frac{\sqrt{2}x}{w|q(z)|}\right) H_m\left(\frac{\sqrt{2}y}{w|q(z)|}\right),
\end{aligned} \tag{6.3}
$$

where $q(z) = 1 + iz/z_0$, $z_0 = kw^2/2$ is the Rayleigh distance, k is the wavenumber of light, (x, y) and (ξ, η) are the transverse Cartesian coordinates in the initial plane and at the distance of z, z is the third Cartesian coordinate directed along the optical axis of the beam. As seen from Eq. (6.3), the HG beams with constant sum of the indices $n + m = N$ propagated with the same phase velocity (i.e. they have the same Gouy phase), and, therefore, the superposition Eq. (6.1) of such beams is propagation-invariant.

Summing the beams Eq. (6.3) with the weight coefficients C_p yields the complex amplitude of the light field Eq. (6.1) at a distance z:

$$SHG_N(x, y, z) = \exp\left[-\frac{x^2 + y^2}{w^2|q(z)|^2} + \frac{ik\left(x^2 + y^2\right)}{2R(z)}\right] \times \exp\left[-i(N + 1)\arctan\left(\frac{z}{z_0}\right)\right] \times \sum_{p=0}^{N} \frac{C_p}{|q(z)|} H_{N-p}\left(\frac{\sqrt{2}x}{w|q(z)|}\right) H_p\left(\frac{\sqrt{2}y}{w|q(z)|}\right), \quad (6.4)$$

with $R(z) = z[1 + (z_0/z)^2]$ being the curvature radius of the wavefront.

Below we derive a general formula for the normalized orbital angular momentum (OAM) of propagation-invariant Gaussian beams Eq. (6.1). To do this, we use the well-known formulae for the OAM J_z of paraxial light fields and for their power W:

$$J_z = \mathrm{Im} \int_{-\infty}^{\infty}\int_{-\infty}^{\infty} E^*(x, y)\left(x\frac{\partial E(x, y)}{\partial y} - y\frac{\partial E(x, y)}{\partial x}\right) dxdy, \quad (6.5)$$

$$W = \int_{-\infty}^{\infty}\int_{-\infty}^{\infty} E^*(x, y)E(x, y)dxdy, \quad (6.6)$$

with $E^*(x, y)$ being the conjugated complex amplitude and Im being the imaginary part of a complex number. Substituting Eq. (6.1) into Eqs. (6.5) and (6.6) as well as by using the orthogonality of the Hermite polynomials and the well-known expression for their derivative

$$\int_{-\infty}^{\infty} \exp\left(-x^2\right)H_n(x)H_m(x)dx = \sqrt{\pi}2^n n!\delta_{n,m},$$
$$\frac{dH_n(x)}{dx} = 2nH_{n-1}(x), \quad n = 1, 2, \ldots N \quad (6.7)$$

with $\delta_{n,m}$ being the Kroneccker delta, we obtain the general formulae for the OAM and power of the beam Eq. (6.1):

$$J_z = \pi 2^N \sum_{p=0}^{N-1} (p + 1)!(N - p)!\mathrm{Im}\left(C_p^* C_{p+1}\right),$$

$$W = \pi 2^{N-1} \sum_{p=0}^{N} p!(N-p)!\left(C_p^* C_p\right). \tag{6.8}$$

The normalized OAM of the field Eq. (6.1) is thus given by

$$\frac{J_z}{W} = \frac{2\sum_{p=0}^{N-1}(p+1)!(N-p)!\mathrm{Im}\left(C_p^* C_{p+1}\right)}{\sum_{p=0}^{N} p!(N-p)!\left(C_p^* C_p\right).}. \tag{6.9}$$

According to Eq. (6.9), the OAM of a propagation-invariant Gaussian beam Eq. (6.1) is nonzero if the coefficients C_p are complex. It is even enough that the coefficients C_p are not all complex, but through one. The minimum sufficient condition for the beam Eq. (6.1) to have nonzero OAM is the phase difference of at least one coefficient in the sum Eq. (6.1) from the phases of other coefficients.

Now we derive particular cases from the general formula (6.9). In a simple case, when only the first two coefficients in the sum Eq. (6.1) are nonzero ($C_0 = 1$, $C_1 = i\gamma$), the OAM [Eq. (6.9)] of such a structured Gaussian beam with the amplitude

$$SHG_N(x, y) = \exp\left(-\frac{x^2 + y^2}{w^2}\right) \times \left[H_N\left(\frac{\sqrt{2}x}{w}\right) + i\gamma 2^{3/2}\frac{y}{w}H_{N-1}\left(\frac{\sqrt{2}x}{w}\right)\right] \tag{6.10}$$

is equal to

$$\frac{J_z}{W} = \frac{2\gamma}{1 + \gamma^2/N}. \tag{6.11}$$

As seen from Eq. (6.11), at large value of $N \gg 1$, the OAM of the beam Eq. (6.10) is determined by the value of γ: $J_z/W \approx 2\gamma$. It can be shown that the topological charge of the field (6.10) is equal to N at any values $\gamma > 0$. Indeed, the amplitude in Eq. (6.10) equals zero when both the real and the imaginary parts of the amplitude are equal to zero. The real part equals zero when the Hermite polynomial $H_N(x)$ is also equal to zero, i.e. in N points, when the argument x is equal to the polynomial's roots. The imaginary part of the amplitude in Eq. (6.10) equals zero at $y = 0$. If $y \neq 0$, then the imaginary part equals zero in the points where the argument x is equal to the roots to the Hermite polynomial $H_{N-1}(x)$. Since the roots to the polynomials $H_N(x)$ and $H_{N-1}(x)$ do not coincide, the field from Eq. (6.10) is equal to zero in the roots of the polynomial $H_N(x)$ and at $y = 0$. There are N such roots and the topological charge of the field from Eq. (6.10) is thus equal to N. For $N = 1$, the expression (10) reduces to an elliptic first-order vortex (topological charge is + 1) with the amplitude $SHG_1(x, y) = \left(2^{3/2}/w\right)\exp\left(-(x^2 + y^2)/w^2\right)(x + i\gamma y)$ and, according to Eq. (6.11), its OAM is equal to $J_z/W = 2\gamma(1 + \gamma^2)^{-1}$. The last expression coincides with one obtained earlier in [17].

Now we consider another simple example of the OAM being proportional to the number $N = 2s + 1$. To do this, we choose only two coefficients in the sum Eq. (6.1) to be nonzero. These are the following two neighboring coefficients:

$$C_{(N-1)/2} = 1, \quad C_{(N+1)/2} = i\gamma, \quad N = 2s + 1. \tag{6.12}$$

Then the complex amplitude of such a superposition of two HG beams is given by

$$SHG_N(x, y) = \exp\left(-\frac{x^2 + y^2}{w^2}\right) \times \left[H_{s+1}\left(\frac{\sqrt{2}x}{w}\right)H_s\left(\frac{\sqrt{2}y}{w}\right) + i\gamma H_s\left(\frac{\sqrt{2}x}{w}\right)H_{s+1}\left(\frac{\sqrt{2}y}{w}\right)\right] \tag{6.13}$$

and its OAM Eq. (6.9) reads as

$$\frac{J_z}{W} = (s + 1)\frac{2\gamma}{1 + \gamma^2} = \left(\frac{N + 1}{2}\right)\frac{2\gamma}{1 + \gamma^2}. \tag{6.14}$$

The vortex beam Eq. (6.13) was studied in [18], while its topological charge was obtained in [19] and is equal to N. Indeed, both terms in Eq. (6.13) are equal to zero simultaneously when the arguments of the Hermite polynomials $H_{s+1}(x)$ in the first term and $H_{s+1}(y)$ in the second term coincide with the roots to the (s+1)th-order Hermite polynomial or when the arguments of the polynomials $H_s(y)$ in the first term and $H_s(x)$ in the second term coincide with the roots to the sth-order Hermite polynomial. Thus, in total, $2s + 1$ zero-intensity points occur, near which the light field from Eq. (6.13) is described by an elliptic first-order optical vortex: $x + i\gamma y$. Therefore, the topological charge of the field from Eq. (6.13) equals $N = 2s + 1$.

Now we consider an example of the coefficients in Eq. (6.1) so that the OAM reaches its maximum value. To do this, we choose the coefficients in Eq. (6.1) being proportional to the binomial coefficients:

$$C_p = \frac{N!(i\gamma)^p}{p!(N - p)!}. \tag{6.15}$$

Then the OAM Eq. (6.9) of the beam Eq. (6.1) with the weight coefficients Eq. (6.15) is equal to

$$\frac{J_z}{W} = N\frac{2\gamma}{1 + \gamma^2}. \tag{6.16}$$

A propagation-invariant Gaussian beam with coefficients Eq. (6.15) can be written in a simpler form using the summation rule of the Hermite polynomials:

$$\begin{aligned} SHG_N(x, y) &= \exp\left(-\frac{x^2+y^2}{w^2}\right) \\ &\times \sum_{p=0}^{N} \frac{N!(i\gamma)^p}{p!(N-p)!} H_{N-p}\left(\frac{\sqrt{2}x}{w}\right) H_p\left(\frac{\sqrt{2}y}{w}\right) \\ &= \exp\left(-\frac{x^2+y^2}{w^2}\right)\left(1-\gamma^2\right)^{N/2} H_N\left(\frac{\sqrt{2}}{w}\frac{x+i\gamma y}{\sqrt{1-\gamma^2}}\right). \end{aligned} \tag{6.17}$$

Beams with the amplitude Eq. (6.17) are called Hermite-Gaussian vortex beams [17]. At $\gamma = 1$, the beam from Eq. (6.17) reduces to a single ringed Laguerre-Gaussian mode with zero radial index ($p = 0$). In this case, all optical vortices with the charge of 1, whose positions are determined by the real-valued zeros of the Hermite polynomial $H_N(x)$, "combine" into a single optical vortex on the optical axis, since at large argument values ($x \gg 1$) the Hermite polynomial is nearly equal to the term with maximal power ($H_N(x) \approx (2x)^N$). The OAM of the beam Eq. (6.17) becomes maximal (16) ($\gamma = 1$) and equal to the OAM of the Laguerre-Gaussian beam $J_z/W = N = l$. The topological charge of the beams Eq. (6.17) was derived in [19] and is equal to N for any value of γ. At $\gamma = 1$, the amplitude of the field from Eq. (6.17) is given by

$$SHG_N(x, y) \approx \frac{\sqrt{2}}{w} \exp\left(-\frac{x^2+y^2}{w^2}\right)(x+i\gamma y)^N. \tag{6.18}$$

We note that in all three above considered examples [Eqs. (6.10), (6.13) and (6.17)], the topological charge is the same and equals N, despite the normalized OAM is different and is given by Eqs. (6.11), (6.14) and (6.16), respectively. Example of non-coinciding topological charge and OAM can be found in [20].

If the coefficients in Eq. (6.1) are chosen as follows:

$$\begin{aligned} C_p &= C_p^{n,m}(\alpha) \\ &= i^p \cos^{n-p}\alpha \sin^{m-p}\alpha P_p^{n-p,m-p}(-\cos 2\alpha), \\ N &= n+m, \end{aligned} \tag{6.19}$$

with $P_p^{\nu,\mu}(x)$ being the Jacobi polynomials, then we obtain the Hermite-Laguerre-Gaussian (HLG) beams, depending by a parameter α and investigated in [21]. We note that for the HLG beams, the number N defines a beams family (rather than a single beam) with different values n and m ($N = m + n$) at a given value of the parameter α. In [21], the normalized OAM has been obtained for the beams with the coefficients Eq. (6.19):

$$\frac{J_z}{W} = (n - m)\sin 2\alpha. \tag{6.20}$$

The indices n and m in Eqs. (6.19) and (6.20) can be related with the azimuthal l and radial p indices of the LG beam: $N = n + m = 2p + l$, $n = p + l$, $m = p$ (see Eq. (6.11) in [16]). Then, if $\alpha = \pi/4$ and $n - m = l$, the HLG beam with the coefficients Eq. (6.19) reduces to the Laguerre-Gaussian beam with the topological charge l and with the maximal OAM equal to $J_z/W = l$.

6.1.2 Numerical Simulation

To confirm that the topological charge of the beams from Eq. (6.10) is equal to N, we computed intensity and phase distributions of such beam for $N = 3$ and $N = 4$ in three different transverse planes. These distributions are illustrated in Fig. 6.1.

As seen in Fig. 6.1, in the initial plane, there are N phase jumps by 2π in the beam periphery. There are also areas where phase changes fast but without the jumps. These areas do not affect the topological charge. Then, on propagation, intensity shape is conserved and each phase distribution contains exactly N first-order optical vortices residing on the horizontal Cartesian axis. Therefore, Fig. 6.1 confirms that the topological charge of the beam from Eq. (6.10) equals N. The numerical OAM value for Fig. 6.1a, d is $J_z/W = 1.4998$, whereas the theoretical value, according to Eq. (6.11), is $J_z/W = 1.5$. For Fig. 6.1g, j, the numerical and theoretical OAM values are 1.5997 and 1.6, respectively.

Figure 6.2 illustrates the intensity and phase distributions of two propagation-invariant superpositions of the HG beams Eq. (6.1) with quasi-random weight coefficients.

As seen in Fig. 6.2, both beams conserve their transverse intensity shape on propagation. The theoretical OAM value Eq. (6.9) for the beam from Fig. 6.2a–f is equal to 1.4452. Numerical OAM value was computed by using the integrals in Eq. (6.5) and Eq. (6.6). The discrete complex amplitude $E(x_i, y_i)$ (256×256 pixels) from Fig. 6.2a was substituted into the integral in Eq. (6.5), which was computed by the trapezium method. As a result, the value J_z was obtained. Similarly, using the integral from Eq. (6.6) for the light field with the complex amplitude $E(x_i, y_i)$, the power W was obtained. Dividing J_z by W yields the numerical OAM value $J_z/W = 1.4451$ for Fig. 6.1a. The OAM values for Figs. 6.2(b) and 2(c) were computed similarly and they turned out to be quite the same ($J_z/W = 1.4451$). Comparison of the OAM value obtained by Eq. (6.9) ($J_z/W = 1.4452$) and numerically ($J_z/W = 1.4451$) indicates that the difference is just in the fourth digit after the decimal point. The theoretical OAM value for the beam from Fig. 6.2g–l is equal to 2.2466. Numerically obtained values are 2.2463 (for all three transverse planes). The topological charge was computed using the M.V. Berry's formula [22]. For Fig. 6.2a–f, $TC = -1$, while for Fig. 6.2g–l it equals $TC = 3$.

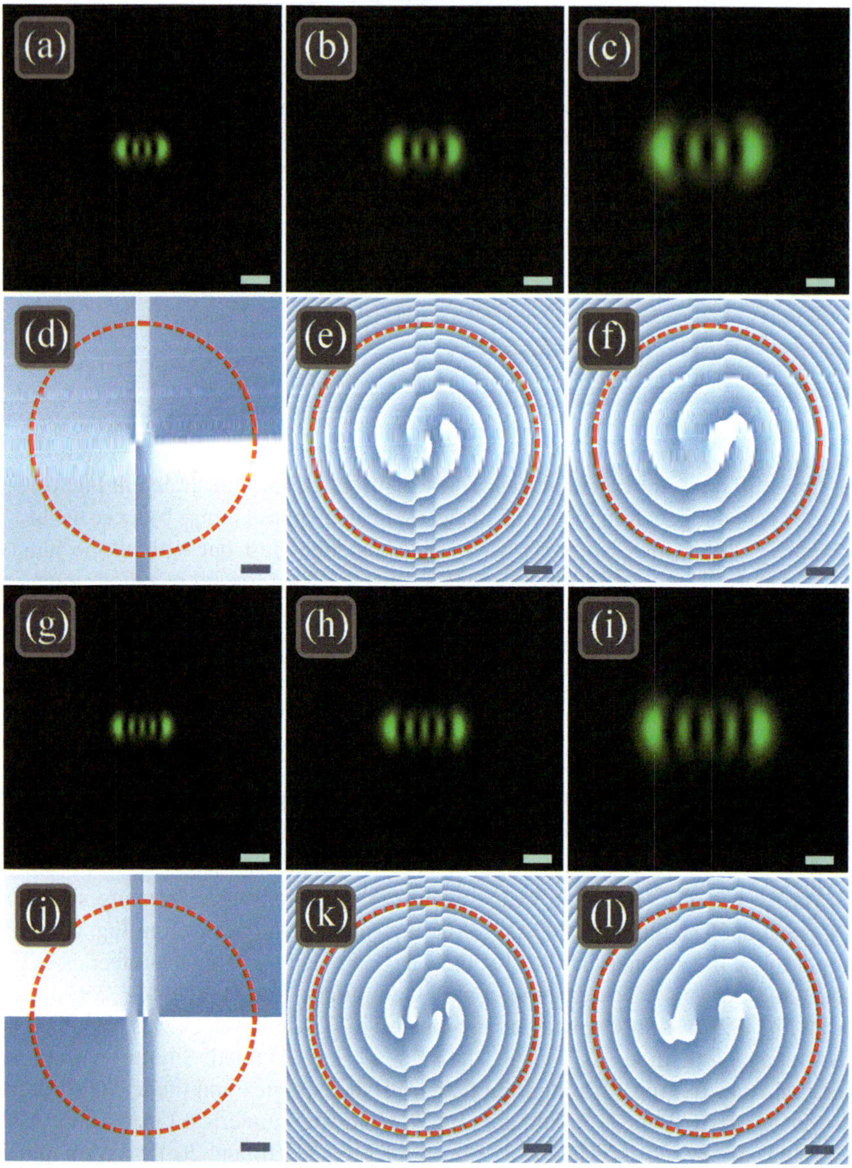

Fig. 6.1 Intensity (**a**–**c**, **g**–**i**) and phase (**d**–**f**, **j**–**l**) distributions of superpositions (6.10) of two Hermite-Gaussian beams for the following parameters: wavelength $\lambda = 532$ nm, Gaussian beam waist radius $w = 0.5$ mm, weight factor $\gamma = 1$, summary order of the Hermite polynomials $N = 3$ (**a**–**f**) and $N = 4$ (**g**–**l**), propagation distances along the optical axis $z = 0$ (initial plane) (**a**, **d**, **g**, **j**), $z = z_0$ (Fresnel diffraction zone) (**b**, **e**, **h**, **k**), and $z = 2z_0$ (far field) (c,f,i,l), The topological charge was calculated along the dashed circle

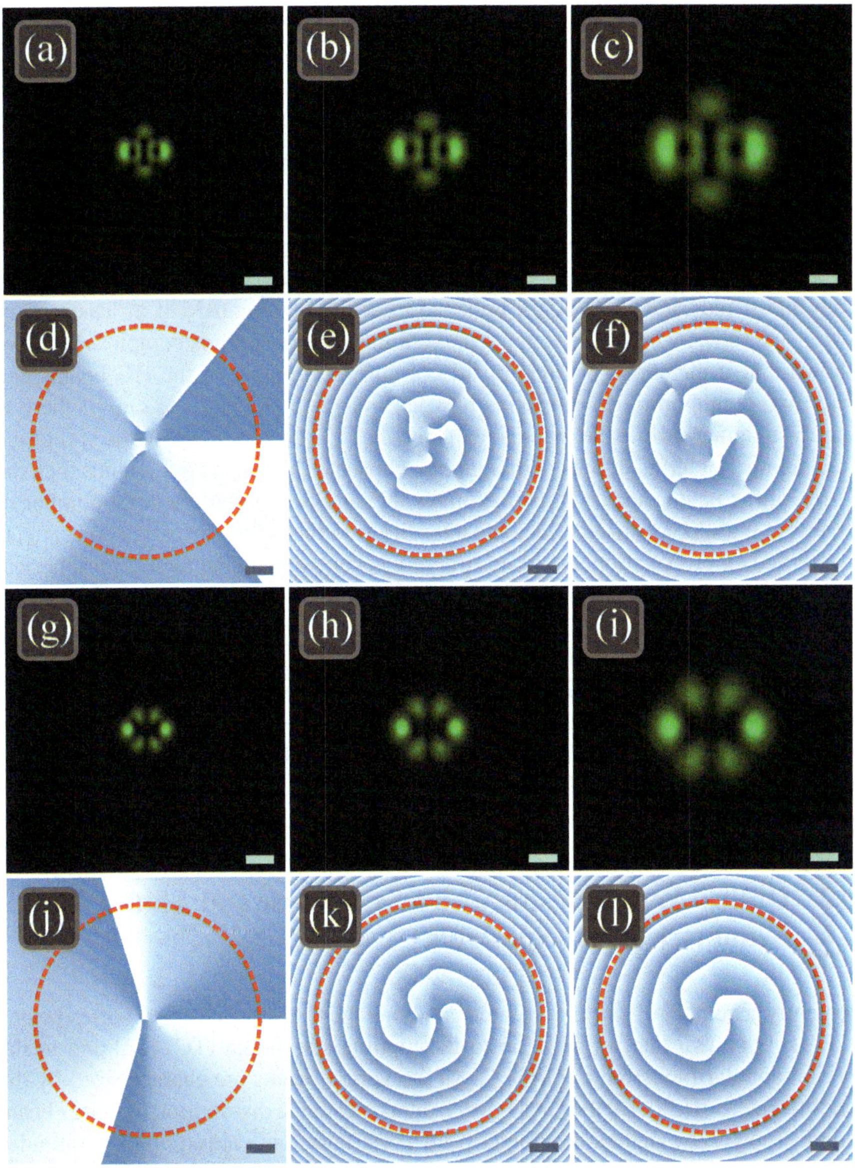

Fig. 6.2 Intensity (**a–c**, **g–i**) and phase (**d–f**, **j–l**) distributions of superpositions of the Hermite-Gaussian beams (6.1) and (6.3) with random weight coefficients for the following parameters: wavelength $\lambda = 532$ nm, Gaussian beam waist radius $w = 0.5$ mm, summary order of the Hermite polynomials $N = 3$, propagation distances along the optical axis $z = 0$ (initial plane) (**a**, **d**, **g**, **j**), $z = z_0$ (Fresnel diffraction zone) (**b**, **e**, **h**, **k**), and $z = 2z_0$ (far field) (**c**, **f**, **i**, **l**), weight coefficients $C_p = [0.77, 0.62i, -0.23, -0.44i]$ (at-f) and $C_p = [0.55, 0.63i, -0.84, -0.07i]$ (**g–l**). The topological charge was calculated along the dashed circle

We note that in random weight coefficients, the phase of each of them differs from the phase of the previous one by $\pi/2$. The phase delay $-\pi/2$ would change the TC to the values $+1$ and -3, respectively.

These numerical examples indicate that the maximal (by modulus) TC of the beam Eq. (6.1) can be equal to the number N or $-N$ from the sum Eq. (6.1). In this case, the normalized OAM is less than N.

We have obtained a general expression for the OAM of a propagation-invariant light field composed of a finite number of the Hermite-Gaussian beams with arbitrary weight coefficients [23]. This expression has been derived in the form of a finite sum and it indicates that such a light field has a nonzero OAM if at least one of two neighboring weight coefficients in the finite superposition is complex. We have shown that if the superposition consists of an even number of the HG beams and only two neighboring central terms have nonzero coefficients, and one of them is purely imaginary while the other is real, then the OAM can be equal to the half of the total number of terms. If the coefficients are chosen as binomial coefficients, then the OAM of the beam can reach its maximal value and be equal to the total number of beams in the superposition. Since the intensity structure of the investigated beams is conserved on propagation, the same structure is generated in the focus of a spherical lens. Tuning the weight coefficients in the superposition of the Hermite-Gaussian beams allows obtaining a given focal-plane intensity distribution and a given OAM value. This advantage of considered here propagation-invariant beams can be used for optical trapping and rotation of microscopic particles [24].

Such beams can be generated by using a phase SLM, with the amplitude-phase function [Eq. (6.1)] being preliminary encoded into phase-only function by using one of the numerous encoding algorithms, for instance, [25].

6.2 Optical Vortex Beams with a Symmetric and Almost Symmetric OAM-Spectrum

Recently, a number of works increased, studying the OAM-spectrum of vortex beams [26]. It is known that, in addition to the topological charge (TC) [27] and to the orbital angular momentum (OAM) [28], the vortex beams are characterized by the OAM-spectrum [29]. By the OAM-spectrum is meant the distribution of vortex beam energy (power) by angular harmonics [29]. These harmonics are complex amplitudes of the form $\exp(in\varphi)$ with n being an integer number and (r, φ) being the polar coordinates in the beam cross-section. For optical decomposition of a light field into the OAM-spectrum, several methods are known, which employ multi-order diffractive optical elements [29–32] or light modulators [33, 34]. Light fields with a wide OAM-spectrum can be generated by using perturbed fork-gratings [35–37], circular diaphragms [38], complicated spiral phase plates [39], or separate holes in an opaque screen [40]. The OAM-spectrum of a light beam can be obtained from an interferogram [35] or simply from a measured beam intensity distribution [36,

37]. Measuring the OAM by using the OAM-spectrum is investigated in many works [41–47]. The OAM of light beams can be measured by using an interferometer [41], by measuring an intensity distribution [43, 47], by using cylindrical lenses [46], or by a single point photodetector [45]. Many methods of sorting the laser modes and optical vortices are based on the decomposition into the OAM-spectrum [48–50].

An optical vortex beam with rotational symmetry has in its transverse cross-section a single n-order intensity null in the center. Among such vortex beams, the well known are the Laguerre-Gaussian modes [50], Bessel-Gaussian beams [51], hypergeometric beams [52, 53], circular beams [54], and the others. TC of such beams is equal to their normalized-to-power OAM and equals n. The OAM-spectrum of such beams consists of a single angular harmonic. A simple shift of the screw dislocation center from the center of the host Gaussian beam allows generation of different OAM-spectrum [46].

An inverse problem is also of interest. Light fields are known with multiple separate screw dislocations distributed over the cross-section of the host Gaussian beam. Examples of these beams are form-invariant beams with their complex amplitude being described by a finite number of isolated intensity nulls [13]. Form-invariant (or structurally stable) beams are coaxial superpositions of the Laguerre-Gaussian optical vortices with the TCs of the same sign. As was proven in [13], such beams conserve their transverse structure on propagation, changing only in scale and rotation. Obtaining the OAM-spectrum of such beams with distributed screw dislocations is equivalent to constructing a polynomial from its roots. A beam with multiple screw dislocations can be obtained from a source rotationally symmetric vortex beam by interfering with a plane wave [55], by elliptical deformation [56], by a complex transverse shift of the complex amplitude function [57], or by an astigmatic transform [6].

In this section, we analyze theoretically a symmetrical OAM-spectrum of optical vortex beams. We show that the order of the central OAM-harmonic of a symmetrical OAM-spectrum is equal to the normalized orbital angular momentum of the light field. In addition, we study two form-invariant beams with an almost symmetrical OAM-spectrum and confirm that, indeed, the order (topological charge) of the average OAM-harmonic in the OAM-spectrum is approximately equal to the normalized-to-power OAM of the beam.

6.2.1 *Orbital Angular Momentum of a Beam with a Symmetric OAM-Spectrum*

Complex amplitude of a paraxial light field in the initial plane can be expanded into an angular-harmonic series:

$$E(r,\varphi) = \sum_{n=-\infty}^{\infty} E_n(r)\exp(in\varphi), \tag{6.21}$$

with (r, φ) being the polar coordinates in the plane $z = 0$. Complex amplitude of each angular harmonic is given by

$$E_n(r) = \frac{1}{2\pi} \int_0^{2\pi} E(r, \varphi) \exp(-in\varphi) d\varphi. \tag{6.22}$$

Each angular harmonic has the following energy (power):

$$W_n = 2\pi \int_0^{\infty} |E_n(r)|^2 r dr, \tag{6.23}$$

while the total energy (power) of the beam is the sum of all the angular harmonic energies:

$$W = \sum_{n=-\infty}^{\infty} W_n. \tag{6.24}$$

The orbital angular momentum of the beam (6.21) can be obtained by the well known expression [58, 59]:

$$\begin{aligned} J_z &= \operatorname{Im} \int_{-\infty}^{\infty} \int_{-\infty}^{\infty} E^*(x, y) \left(x \frac{\partial}{\partial y} - y \frac{\partial}{\partial x} \right) E(x, y) dx dy \\ &= \operatorname{Im} \int_0^{\infty} \int_0^{2\pi} E^*(r, \varphi) \frac{\partial E(r, \varphi)}{\partial \varphi} r dr d\varphi, \end{aligned} \tag{6.25}$$

where Im is the imaginary part of a complex number, E^* means the complex conjugation. Substituting Eq. (6.21) into Eq. (6.25) and by using Eq. (6.23), we obtain a well known expression for the OAM of an arbitrary paraxial beam [60]:

$$J_z = \sum_{n=-\infty}^{\infty} n W_n. \tag{6.26}$$

In this work, we obtain some general properties of a symmetric OAM-spectrum and study several examples. We suppose that the powers of the angular harmonics (OAM-spectrum) are distributed symmetrically with respect to some average harmonic of the order n_0, i.e.

$$W_{n_0+n} = W_{n_0-n}. \tag{6.27}$$

For such a symmetric OAM-spectrum, order (TC) of the average angular harmonic equals the OAM of the whole beam, normalized to its power. Indeed,

$$\begin{aligned} J_z &= \sum_{n=-\infty}^{\infty} nW_n = n_0 W_{n_0} \\ &+ \sum_{n=1}^{\infty} \left[(n_0 + n) W_{n_0+n} + (n_0 - n) W_{n_0-n} \right] \\ &= n_0 W_{n_0} + 2n_0 \sum_{n=1}^{\infty} W_{n_0+n} = n_0 W. \end{aligned} \tag{6.28}$$

Equation (6.28) is derived by using Eqs. (6.24) and (6.27). This equation means that the normalized-to-power OAM of the beam with a symmetrical OAM-spectrum is equal to the TC of the average angular harmonic. According to Eq. (6.28), all light fields with a symmetrical OAM-spectrum and with the same order of the central harmonic have the same normalized-to-power OAM, equal to the order of this central angular harmonic. Thus, Eq. (6.28) leads to yet another method for changing the shape of a light field by changing the width and shape of its OAM-spectrum, but with the beam OAM remained unchanged. Other methods of varying the light field amplitude with the OAM remaining unchanged are also known [61].

6.2.2 Sample Beam with Symmetric OAM-Spectrum

Complex amplitude of a light field with the symmetric OAM-spectrum can be obtained by shifting the OAM-spectrum, symmetric with respect to the 0th angular harmonic:

$$E_s(r, \varphi) = \sum_{n=-\infty}^{\infty} E_{|n|}(r) \exp(i(n + n_0)\varphi). \tag{6.29}$$

The OAM of the beam (6.29) can be shown to equal $J_z = n_0 W$. Now we show that the field (6.29) has the OAM-spectrum, symmetric with respect to the central angular harmonic n_0. If $m = n_0 \pm n$, then the amplitude E_m of the mth OAM-harmonic of the field (6.29) is equal (the argument r is omitted for brevity):

$$\begin{aligned} E_m &= \frac{1}{2\pi} \int_0^{2\pi} E_s(r, \varphi) \exp(-im\varphi) d\varphi \\ &= \frac{1}{2\pi} \int_0^{2\pi} \sum_{p=-\infty}^{\infty} E_{|p|}(r) \exp\left[i(p + n_0 - m)\varphi\right] d\varphi \end{aligned}$$

$$= E_{|m-n_0|} = E_{|n_0+n-n_0|} = E_{|n|} = E_{|n_0-n-n_0|} = E_{|-n|}. \tag{6.30}$$

For instance, we can consider a simple light field with the symmetric OAM-spectrum:

$$E_s(r, \varphi) = \exp\left(-\frac{r^2}{w^2}\right) \times \sum_{n=-\infty}^{\infty} C_n \left(\frac{\sqrt{2}r}{w}\right)^{|n|} \exp[i(n + n_0)\varphi], \tag{6.31}$$

with w being the waist radius of the Gaussian beam and C_n being the superposition coefficients. For example, if $C_n = 1/|n|!$, each OAM harmonic of the field (6.31) has the following energy (6.23).

$$W_n = \frac{\pi w^2}{2|n|!}. \tag{6.32}$$

In this case, the total power (6.24) of the beam is $W = \pi w^2(\mathrm{e} - 1/2)$. The order n_0 of the central OAM-harmonic of the beam (6.31) is coinciding exactly with its the normalized-to-power OAM $J_z/W = n_0$. Unfortunately, the beam (6.31) is not a mode and changes its transverse intensity shape on propagation, since its radial and azimuthal factors have different indices. In the Fresnel diffraction zone, each term in the sum (6.31) is proportional to the confluent hypergeometric function ${}_1F_1(a, b, \xi)$ [52, 53].

6.2.3 *Family of Form-Invariant Beams with an Almost Symmetric OAM-Spectrum*

As another example, we consider here a family of form-invariant light fields with the complex amplitude of the following form:

$$E_{ss}(r, \varphi) = \exp\left(-\frac{r^2}{w^2}\right) f\left(re^{i\varphi}\right) = \exp\left(-\frac{r^2}{w^2}\right) \sum_{n=0}^{\infty} \frac{C_n r^n}{n!} \exp(in\varphi). \tag{6.33}$$

In Eq. (6.33), $C_n = \mathrm{d}^n f(z)/\mathrm{d}z^n$ at $z = 0$. The function $f(re^{i\varphi})$ in Eq. (6.33) is an entire function and, therefore, can be expanded into a converging Taylor series. As seen in Eq. (6.33), the OAM-spectrum of the form-invariant field is positive, i.e. all angular harmonics have a nonnegative TC ($n \geq 0$). Such light fields preserve their transverse structure on propagation, only changing in scale and rotating around the optical axis. However, the OAM-spectrum of the beams (6.33) is not symmetric. Only for some fields, almost symmetric spectrum can be obtained. For example, if we consider a function $f(re^{i\varphi}) = \cos(re^{i\varphi}/\alpha)$ (with α being some constant), then instead of Eq. (6.33) we get

$$E_c(r,\varphi) = \exp\left(-\frac{r^2}{w^2}\right)\cos\left(\frac{r}{\alpha}e^{i\varphi}\right)$$
$$= \exp\left(-\frac{r^2}{w^2}\right)\sum_{p=0}^{\infty}\frac{(-1)^p}{(2p)!}\left(\frac{r}{\alpha}\right)^{2p}\exp(2ip\varphi). \tag{6.34}$$

It is seen from Eq. (6.34) that the OAM-spectrum contains only even nonnegative angular harmonics. From Eq. (6.34), the energy of each OAM-harmonic is given by

$$W_{2p} = \frac{\pi w^2}{2(2p)!}\left(\frac{w^2}{2\alpha^2}\right)^{2p}. \tag{6.35}$$

The total (summary) energy of the beam (6.34) is

$$W = \sum_{p=0}^{\infty} W_{2p} = \frac{\pi w^2}{2}\cosh\left(\frac{w^2}{2\alpha^2}\right). \tag{6.36}$$

The orbital angular momentum of the beam (6.34) is equal to

$$J_z = \sum_{n=-\infty}^{\infty} nW_n = \left(\frac{\pi w^2}{2}\right)\left(\frac{w^2}{2\alpha^2}\right)\sinh\left(\frac{w^2}{2\alpha^2}\right), \tag{6.37}$$

and the normalized-to-power OAM is thus given by

$$\frac{J_z}{W} = \left(\frac{w^2}{2\alpha^2}\right)\tanh\left(\frac{w^2}{2\alpha^2}\right). \tag{6.38}$$

If the cosine period in the complex amplitude (6.34) is small, i.e. $w^2 \gg 2\alpha^2$, the hyperbolic tangent in Eq. (6.38) is approximately equal to 1. Therefore, the OAM is nearly equal to $J_z/W \approx w^2/(2\alpha^2) \gg 1$. Now we show that if $w^2 \gg 2\alpha^2$, the order (TC) of the OAM-harmonic contributing the most into the OAM-spectrum of the beam (6.34), is equal to the normalized-to-power OAM of this beam. Indeed, if the cosine period is small (i.e. $w^2 \gg 2\alpha^2$), the energy (6.35) of each OAM-harmonic (6.35) is proportional to an expression $W_n \sim \xi^n/n!$ (with $\xi = w^2/(2\alpha^2)$), which achieves its maximal value, according to the Stirling's formula ($n! \approx n^n$), at $n = \xi$.

The same considerations can also be done for another example, for comparison. If we consider a function $f(re^{i\varphi}) = \cos^2(re^{i\varphi}/\alpha)$, then, instead of Eq. (6.34), we get

$$
\begin{aligned}
E_c(r,\varphi) &= \exp\left(-\frac{r^2}{w^2}\right)\cos^2\left(\frac{r}{\alpha}e^{i\varphi}\right) \\
&= \exp\left(-\frac{r^2}{w^2}\right)\sum_{p=0}^{\infty}\frac{\theta_n(-1)^p}{2(2p)!}\left(\frac{2r}{\alpha}\right)^{2p} e^{2ip\varphi},
\end{aligned}
\tag{6.39}
$$

with $\theta_n = \{1,\ n > 1;\ 2,\ n = 0\}$. Expression (6.39) differs from (6.34) since the cosine function is squared. Light fields described by Eq. (6.39) or by Eq. (6.34) are the examples of optical vortex beams with an almost symmetric OAM-spectrum. The OAM-spectrum is symmetric only approximately, since only OAM-harmonics of nonnegative even orders are nonzero

It is seen from Eq. (6.39) that the OAM-spectrum has only even nonnegative angular harmonics. The energy of each OAM-harmonic in Eq. (6.39) is given by

$$
\begin{aligned}
W_{2p} &= \frac{\pi w^2}{8(2p)!}\left(\frac{2w^2}{\alpha^2}\right)^{2p}, \quad p > 0, \\
W_0 &= \frac{\pi w^2}{2}, \quad p = 0.
\end{aligned}
\tag{6.40}
$$

If $2w^2 > \alpha^2$, then with increasing p the energy of OAM-harmonics (6.40) at first grows, but then decays (because of the factorial in the denominator). Since, according to the Stirling's formula, the factorial $n!$ at large n is approximately equal to n^n, the energy (6.40) is maximal when the numerator and denominator are nearly equal (i.e. $W_n \sim \eta^n/n! \approx (\eta/n)^n = 1$, where $\eta = 2w^2/\alpha^2$):

$$
\frac{2w^2}{\alpha^2} \approx 2p. \tag{6.41}
$$

Total energy of the beam (6.39) is equal to

$$
W = \sum_{n=-\infty}^{\infty} W_n = \frac{3\pi w^2}{8} + \left(\frac{\pi w^2}{8}\right)\cosh\left(\frac{2w^2}{\alpha^2}\right), \tag{6.42}
$$

while the beam's OAM is

$$
J_z = \sum_{n=-\infty}^{\infty} nW_n = \left(\frac{\pi w^2}{8}\right)\left(\frac{2w^2}{\alpha^2}\right)\sinh\left(\frac{2w^2}{\alpha^2}\right). \tag{6.43}
$$

If the cosine period is small, i.e. $2w^2 \gg \alpha^2$, the normalized-to-power OAM of the beam (6.39) equals

$$
\frac{J_z}{W} = \frac{\eta\sinh(\eta)}{3+\cosh(\eta)} \approx \eta = \frac{2w^2}{\alpha^2}. \tag{6.44}
$$

The OAM-spectra (6.35) and (6.40) of the light fields (6.34) and (6.39) are symmetric only approximately, and the order of the central harmonic ($2w^2/\alpha^2 \approx 2p$), which contributes the most into the OAM-spectrum, coincides with the approximate expression for the normalized-to-power OAM: $J_z/W \approx 2w^2/\alpha^2 >> 1$.

Thus, we have shown that the light fields (6.29) with the symmetric OAM-spectrum do not conserve their shape on space propagation, while the form-invariant fields (6.34) and (6.39) have the OAM-spectrum, only approximately symmetric. However, in all three given examples [Eqs. (6.31), (6.34) and (6.39)], the order (TC) of the average OAM-harmonic is coinciding (either exactly or approximately) with the normalized-to-power OAM of the vortex beam. We note that the beams (6.34) and (6.39) have a countable number of screw dislocations, each with the TC = 1 [for beam (6.34)] or TC = 2 [for beam (6.39)]. Therefore, both beams have the infinite TC. Thus, it seems that the beams cannot be compared by their TC. However, the beams (6.34) and (6.39) can be compared by their OAM. For instance, the normalized OAM of the beam (6.39) ($J_z/W \approx 2w^2/\alpha^2$) is four times greater than that of the beam (6.34) ($J_z/W \approx w^2/(2\alpha^2)$). Since the central OAM-harmonic of the OAM-spectrum of the beam (6.39) has the TC four times higher than the central OAM-harmonic of the beam (6.34), the TC of the central OAM-harmonic of the beam (6.39) is also four times greater than that of the beam (6.34).

6.2.4 Numerical Simulation

Simulation of beams with a symmetric OAM-spectrum

In this section, we consider as an example a light beam (6.31) with normally distributed contributions of the OAM-harmonics, i.e. $C_n = \exp[-n^2/(2\sigma^2)]$. In the initial plane, complex amplitude of such beam reads as

$$E_s(r,\varphi) = \exp\left(-\frac{r^2}{w^2}\right) \times \sum_{n=-\infty}^{\infty} \exp\left(-\frac{n^2}{2\sigma^2}\right)\left(\frac{\sqrt{2}r}{w}\right)^{|n|} \exp[i(n+n_0)\varphi], \tag{6.45}$$

where σ defines the decay ratio of the OAM-harmonics contributions from the central n_0th harmonic to the periphery.

Figure 6.3 depicts the OAM-spectra, as well as the intensity and phase distributions of the beam (6.45) in the initial plane and at the Rayleigh distance, for several symmetrical OAM-spectra with the 10th-order central harmonic and with different decaying ratios of secondary (noncentral) harmonics contributions.

In the initial plane, maximal intensity of each harmonic in Eq. (6.45) is on a ring with a radius $r_{\max,n} = w(|n|/2)^{1/2}$ and is equal to $I_{\max,n} = |n|^{|n|}\exp(-|n| - n^2/\sigma^2)$. Therefore, even at the slowest decaying of the angular harmonics contributions in

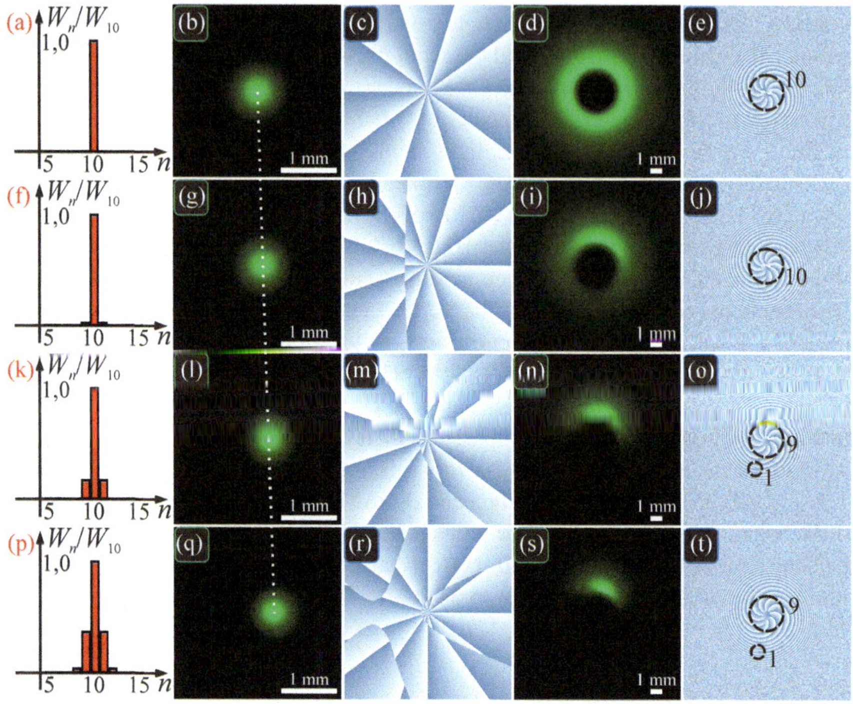

Fig. 6.3 OAM-spectra (**a**, **f**, **k**, **p**), as well as distributions of the normalized-to-maximum intensity (**b**, **d**, **g**, **i**, **l**, **n**, **q**, **s**) and phase (**c**, **e**, **h**, **j**, **m**, **o**, **r**, **t**) in the initial plane (**b**, **c**, **g**, **h**, **l**, **m**, **q**, **r**) and at the Rayleigh distance (**d**, **e**, **i**, **j**, **n**, **o**, **s**, **t**) of the beam (6.45) for the following parameters: wavelength $\lambda = 532$ nm, Gaussian beam waist radius $w = 0.5$ mm, order of the central OAM-harmonic $n_0 = 10$, decaying ratio of the OAM-harmonics contributions $\sigma = 0.25$ (a-e), $\sigma = 0.5$ (f-j), $\sigma = 0.75$ (k–o), $\sigma = 1.00$ (p–t), computation domain $|x|, |y| \leq R$ with $R = 1.5$ mm (intensity in the initial plane), $R = 5$ mm (phase in the initial plane), and $R = 7.5$ mm (intensity and phase at the Rayleigh distance). The dashed circles on the phase distributions at the Rayleigh distance (**e**, **j**, **o**, **t**) show the area with optical vortices (their number is also written near the circles). The white dotted line (**b**, **g**, **l**, **q**) shows the shift of the Gaussian beam center

Fig. 6.3, i.e. at $\sigma = 1$, maximal intensity of the $\pm$ 2nd harmonics is only 1% of the maximal intensity of the 0th harmonic, while the intensity of higher-order harmonics is even lower. Therefore, the intensity distribution in the initial plane consists mainly from the zero harmonic and is slightly distorted by $\pm$ 1st-order harmonics. As a result, the initial pattern for all four values σ has a Gaussian shape, which slowly moves to the right with increasing σ. For instance, at $\sigma = 0.25$, the superposition (6.45) is contributed mostly from the following three terms:

$$E_s(r, \varphi) = \exp\left(-\frac{r^2}{w^2}\right) \exp(i10\varphi)$$

$$+\exp(-8)\exp\left(-\frac{r^2}{w^2}\right)\left(\frac{\sqrt{2}r}{w}\right)\exp(i11\varphi)$$

$$+\exp(-8)\exp\left(-\frac{r^2}{w^2}\right)\left(\frac{\sqrt{2}r}{w}\right)\exp(i9\varphi)+\ldots \quad (6.46)$$

It is seen from Eq. (6.46) that the second and the third terms have much weaker amplitude than the first term. Therefore, in the initial plane, the vortex beam (6.45) has a shape of the Gaussian beam (Fig. 6.3b). The beam (6.46) has the TC of 10 (Fig. 6.3c), since for an arbitrary value of r the first term of the series (6.46) overwhelms all the rest terms. However, since the high-TC optical vortex is unstable, then, due to the other terms in (6.46) in addition to the main first term, the central 10th-order vortex splits on propagation into 10 optical vortices with TC of + 1, located close to each other (Fig. 6.3e). We note that the separate vortices in the center of the phase distribution (Fig. 6.3e) are not the additional vortices occurring due to the spin–orbit interaction [62]. We limit our study here by the paraxial scalar approximation and the beams do not have circular polarization as in [62].

The beams in Eq. (6.45) are not modes and, on space propagation, other, noncentral, harmonics become more eminent. When σ is small, the OAM-spectrum is very narrow. Therefore, at $\sigma = 0.25$, the intensity distribution at the Rayleight distance is rotationally symmetric (Fig. 6.3d). When $\sigma = 0.5$, the OAM-spectrum is visually almost unchanged, but the intensity distribution acquires a shape of an asymmetric light ring (Fig. 6.3i), which, with increasing contributions of secondary harmonics (i.e. with increasing σ), is broken up and becomes a crescent (Fig. 6.3n, s). Despite the distorted intensity shape, the normalized OAM, according to Eq. (6.28), should be equal to $n_0 = 10$ in all cases. Indeed, in all four phase distributions (Fig. 6.3e, j, o, t), 10 optical vortices can be seen (marked by dashed circles in Fig. 6.3), one of which moves away from the center with increasing σ. The normalized-to-power OAM, computed by Eq. (6.25) and divided by the beam power, also has the values close to 10: 9.35 (Fig. 6.3b, c), 9.46 (Fig. 6.3d, e), 9.37 (Fig. 6.3g, h), 9.43 (Fig. 6.3i, j), 9.48 (Fig. 6.3l, m), 9.45 (Fig. 6.3n, o), 9.58 (Fig. 6.3q, r), 9.50 (Fig. 6.3s, t).

The shift of the Gaussian beam (Fig. 6.3b, g, l, q) in the initial plane to the right along the x-axis with increasing σ can be explained by Eq. (6.46), which can be rewritten as

$$E_s(r,\varphi)=\exp\left(-\frac{r^2}{w^2}\right)\exp(i10\varphi)\left(1+\frac{\alpha r}{w}\cos\varphi\right)+\cdots \quad (6.47)$$

with $\alpha = 2^{3/2}\exp[-1/(2\sigma^2)]$. According to Eq. (6.47), the intensity maximum shifts along the axis $x = r\cos\varphi$ with increasing α:

$$I(r,\varphi)=|E_s(r,\varphi)|^2\approx\exp\left(-\frac{2r^2}{w^2}+\frac{2\alpha x}{w}\right)+\cdots \quad (6.48)$$

Equation (6.47) also demonstrates that near $x \approx -w/\alpha$, the amplitude is zero, i.e. there is a phase jump by π (edge dislocation) along the vertical line $x \approx -w/\alpha$. This phase jump is clearly seen in Fig. 6.3h. With further increase of α, contribution of the other terms of the series (6.45) into Eq. (6.47) is also increasing:

$$E_s(r, \varphi) = \exp\left(-\frac{r^2}{w^2}\right) \exp(i10\varphi) \times \left(1 + \frac{\alpha r}{w} \cos\varphi + \frac{\beta r^2}{w^2} \cos 2\varphi\right) + \cdots \tag{6.49}$$

with $\beta = 4\exp(-2/\sigma^2)$. According to Eq. (6.49), the line of zero amplitude, where the phase has a jump by π (Fig. 6.3m), is now not a straight line as in Fig. 6.3h, but a curve (Fig. 6.3m). The separation of one vortex with the TC + 1 from the 10 vortices in the center (Fig. 6.3o, t) can be explained by interaction between the edge dislocation with an optical vortex (screw dislocation) (Fig. 6.3m, r) [63, 64]. The difference from [64, 65] is that, instead of the straight edge dislocation, it is bended in Fig. 6.1m, r. In order to give a detailed explanation of this phenomenon, the Fresnel transform of the function (6.49) should be obtained. This would lead to rather cumbersome expressions including the difference of two Bessel functions of half-integer orders [65], analysis of which would lead us far from the goal of this work.

In conclusion of this section, we give a general formula for the field (6.45), similar to (6.49), which clearly illustrates that both the TC and the normalized-to-power OAM of the beam with a symmetric OAM-spectrum is equal to n_0:

$$E_s(r, \varphi) = \exp\left(-\frac{r^2}{w^2}\right) \exp(in_0\varphi) \times \left[1 + \sum_{n=1}^{\infty} \frac{c_n r^n \cos n\varphi}{w^n}\right], \tag{6.50}$$

where

$$c_n = 2^{(|n|+2)/2} \exp\left(-\frac{n^2}{2\sigma^2}\right). \tag{6.51}$$

As seen from Eq. (6.50), since the sum is a real-valued, the TC of the field (6.50) is equal to n_0 and the OAM (6.25), normalized to the beam power (6.24), is also equal to $J_z/W = n_0$. In this work, we do not consider the intrinsic and extrinsic components of the OAM (6.25), and in all cases we calculate it relative to the origin. Of course, in the cases when the ‘center of mass’ of the beam intensity distribution is shifted from the optical axis (i.e. the beam has an asymmetric intensity distribution), the OAM is a sum of two terms, only one of which is the intrinsic OAM.

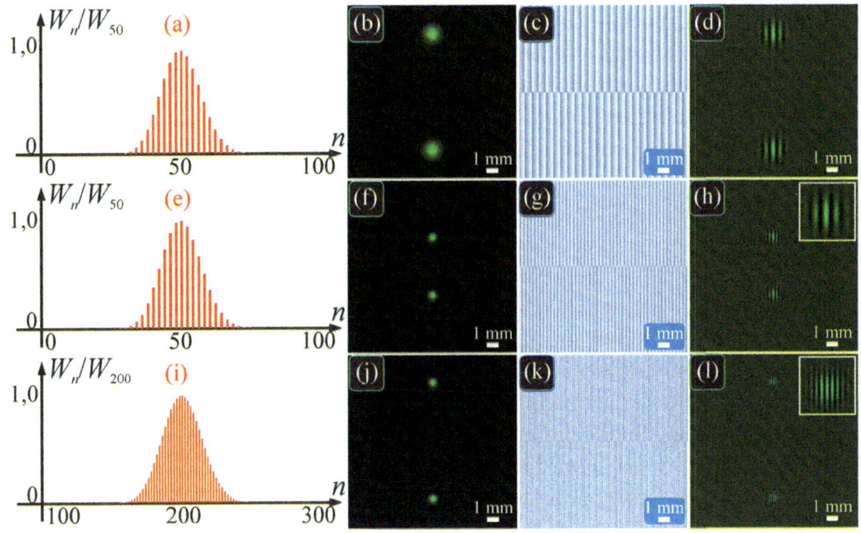

Fig. 6.4 OAM-spectra (**a**, **e**, **i**), normalized-to-maximum intensity (**b**, **f**, **j**) and phase (**c**, **g**, **k**) distributions, as well as intensity distributions with an interfering plane wave (**d**, **h**, **l**) of the beams (6.34) (**a**–**h**) and (6.39) (**i**–**l**) in the initial plane for the following parameters: wavelength $\lambda = 532$ nm, Gaussian beam waist radius $w = 1$ mm (**a**–**d**) and $w = 0.5$ mm (**e**–**l**), scaling factor of the vortex cosine function $\alpha = w/10$, computation domain $|x|, |y| \leq R$ with $R = 7.5$ mm. The insets (**h**, **l**) show the $3 \times$ magnified fragments of the interference patterns

Numerical simulation of form-invariant beams with an almost symmetric oam-spectrum

In this Section, we investigate numerically the form-invariant light beams (6.34) and (6.39). Figure 6.4 shows the OAM-spectra, as well as the intensity and phase distributions of these beams in the initial plane.

Figures 6.4a–d and e–h show the beams (6.34) with the different waist radius w, but with the same ratio $w/\alpha = 10$. Therefore, according to Eq. (6.38), their OAM-spectra should be the same (up to a constant multiplier) and, indeed, the normalized-to-maximum OAM-spectra in Fig. 6.4a, e look similar. The theory predicts that the normalized OAM equals $J_z/W \approx w^2/(2\alpha^2) = 50$ in both cases. The OAM values obtained numerically using Eq. (6.25) are 49.82 (Fig. 6.4a–d) and 49.27 (Fig. 6.4e–h). Despite the different shape of the beams (6.34) at $w = 1$ mm (Fig. 6.4a–d) and $w = 0.5$ mm (Fig. 6.4e–h), their effective TC is the same and the number of stripes in the interference patterns is also the same (in Fig. 6.4d, h, five stripes are clearly seen). The equal OAMs of the beams from Fig. 6.4a–d and e–h can be physically explained by an analogy with the mechanical torque.

Compared to the beam with $w = 0.5$ mm (Fig. 6.4e–h), the beam with $w = 1$ mm (Fig. 6.4a–d) has a twice greater leverage (the distance from the center of the pattern to the light spots), but two times lower transverse force (velocity of phase changes in the transverse plane). According to the theory, the beam (6.39) (Fig. 6.4i–l) should have the maximal-energy OAM-harmonic with the TC four times higher, and the normalized OAM should also increase four times. Numerical calculation gives the OAM value 188.66, while the largest contribution is given by 200th OAM-harmonic (Fig. 6.4i). The interference pattern (Fig. 6.4l) has two times more interference stripes (clearly seen 9 stripes), but, at the same time, the light spots themselves in Fig. 6.4j are also two times further from the center compared to Fig. 6.4f, i.e., by analogy with the mechanical torque, both the leverage and the force increase two times. The analogy between the transverse force, acting onto a particle in the light field, and the velocity of the phase change has the following explanation. The force is proportional to the transverse energy flow, which is proportional to the energy (intensity) density in the current point multiplied by the phase gradient. The greater is the phase gradient, the greater is the energy flow and the force. In [66], the energy flow expressed via the phase gradient of a scalar field, is called an optical current. That is why the OAM increases four times. Similarly, it can be shown that the effective TC for the case from Fig. 6.4i–l is four times greater than that for the case from Fig. 6.4e–h. Indeed, for a conventional (circular) optical vortex with the phase $n\varphi$, the TC equals the length of a circumference with the radius R divided by a period T of phase change by 2π: $TC = 2\pi R/T$. And since the period T of the phase change by 2π in Fig. 6.4l is two times smaller than that in Fig. 6.4h (i.e. two times more interference stripes), and the distance R to the center of the Gaussian beam in Fig. 6.4l is two times greater than in Fig. 6.4h, then the effective TC of the beam in Fig. 6.4l is four times greater than TC of the beam from Fig. 6.4h.

Thus, we have shown theoretically and confirmed numerically that if a vortex beam has a symmetrical OAM-spectrum, then the TC of the central OAM-harmonic coincides with the beam's the normalized-to-power OAM [67]. This general property of the vortex beams is also valid for beams with an approximately (almost) symmetrical OAM-spectrum. For two examples of form-invariant beams with an infinite TC and with a finite OAM, we have demonstrated that they have almost symmetrical OAM-spectrum and that the TC of the central angular harmonic is almost equal to the tnormalized-to-power OAM of these beams. Therefore, varying the width of the symmetrical OAM-spectrum without changing the central OAM-harmonic (i.e. without changing its position), vortex beams can be obtained with different intensity distributions, but with the same OAM value. This is yet another method of changing the beam shape without changing its 'effective' TC and orbital angular momentum. In practice, laser beams with the symmetric OAM-spectrum can be generated by using either a SLM or diffractive optical elements [25, 68].

References

1. A. Forbes, Structured light from lasers. Las. Phot. Rev. **13**, 1900140 (2012)
2. J. Wang, Y. Liang, Generation and detection of structured light: a review. Front. Phys. **9**, 688284 (2021)
3. Y. Bai, H. Lv, X. Fu, Y. Yang, Vortex beam: generation and detection of orbital angular momentum. Chin. Opt. Lett. **20**, 012601 (2022)

4. S. Scholes, H. Sroor, K. Ait-Ameur, Q. Zhan, A. Forbes, General design principle for structured light lasers. Opt. Express **28**, 35006–35017 (2020)
5. J. Pan, Y. Shen, Z. Wan, X. Fu, H. Zhang, Q. Liu, Index-tunable structured light beams from a laser with an intracavity astigmatic mode converter. Phys. Rev. Appl. **14**, 044048 (2020)
6. E.G. Abramochkin, V.G. Volostnikov, Beam transformations and nontransformed beams. Opt. Commun.Commun. **83**, 123–135 (1991)
7. S. Restuccia, D. Giovannini, G. Gibson, M.J. Padgett, Comparing the information capacity of Laguerre-Gaussian and Hermite-Gaussian modal set in a finite-aperture system. Opt. Express **24**, 27127–27136 (2016)
8. G.A. Siviloglou, J. Broky, A. Dogariu, D.N. Christodoulides, Observation of accelerating airy beams. Phys. Rev. Lett. **99**, 213901 (2007)
9. Q.W. Zhan, Cylindrical vector beams: from mathematical concepts to applications. Adv. Opt. Photon. **1**, 1–57 (2009)
10. C. Chong, J. Wan, Q.W. Chen, Zhan, Generation of spatiotemporal optical vortices with controllable transverse orbital angular momentum. Nat. Photon. **14**, 350–354 (2020)
11. L. Rego, K.M. Dorney, N.J. Brooks, Q.L. Nguyen, C.T. Liao, J.S. Roman, D.E. Couch, A. Liu, E. Pisanty, M. Lewenstein, L. Plaja, H.C. Kapteyn, M.M. Murnane, C. Hernández-García, Generation of extreme-ultraviolet beams with time-varying orbital angular momentum. Science **364**, eaaw9486 (2019)
12. G. Indebetouw, Optical vortices and their propagation. J. Mod. Opt. **40**, 73–87 (1993)
13. E.G. Abramochkin, V.G. Volostnikov, Spiral-type beams: optical and quantum aspects. Opt. Commun.Commun. **125**, 302–323 (1996)
14. V.V. Kotlyar, Optical beams with an infinite number of vortices. Computer Opt. **45**, 490–496 (2021)
15. M. Soskind, R. Soskind, Y. Soskind, Shaping propagation invariant laser beams. Opt. Engin. **54**, 111309 (2015)
16. V. Volyar, E. Abramochkin, Yu. Egorov, M. Bretsko, Ya. Akimova, Fine structure of perturbed Laguerre-Gaussian beams: Hermite-Gaussian mode spectra and topological charge. Appl. Opt. **59**, 7680–7687 (2020)
17. V.V. Kotlyar, A.A. Kovalev, A.P. Porfirev, Vortex Hermite-Gaussian laser beams. Opt. Lett. **40**, 701–704 (2015)
18. V.V. Kotlyar, A.A. Kovalev, Hermite-Gaussian modal laser beams with orbital angular momentum. J. Opt. Soc. Am. A **31**, 274–282 (2014)
19. V.V. Kotlyar, A.A. Kovalev, Topological charge of asymmetric optical vortices. Opt. Express **28**, 20449–20460 (2020)
20. V.P. Aksenov, V.V. Dudorov, G.A. Filimonov, V.V. Kolosov, V.Y. Venediktov, Vortex beams with zero orbital angular momentum and non-zero topological charge. Opt. Las. Technol. **104**, 159–163 (2018)
21. E.G. Abramochkin, V.G. Volostnikov, Generalized Gaussian beams. J. Opt. A Pure Appl. Opt. **6**, S157–S161 (2004)
22. Pure and Appl, M.V. Berry, Optical vortices evolving from helicoidal integer and fractional phase steps. J. Opt. A Pure Appl. Opt. **6**, 259–268 (2004)
23. V.V. Kotlyar, A.A. Kovalev, Orbital angular momentum of paraxial propagation-invariant laser beams. J. Opt. Soc. Am. A **39**, 1061–1065 (2022)
24. Y. Yang, Y. Ren, M. Chen, Y. Arita, C. Rosales-Guzman, Optical trapping with structured light: a review. Adv. Phot. **3**, 034001 (2021)
25. O. Mendoza-Yero, G. Mínguez-Vega, J. Lancis, Encoding complex fields by using a phase-only optical element. Opt. Lett. **39**, 1740–1743 (2014)
26. V.V. Kotlyar, A.A. Kovalev, A.P. Porfirev, *Vortex laser beams* (CRC Press, Boca Raton, 2018)
27. V.V. Kotlyar, A.G. Nalimov, Evolution of a fractional-charge optical vortex upon free-space propagation. Optik **261**, 169158 (2022)
28. L. Allen, M.V. Beijerbergen, R.J.C. Spreeuw, J.P. Woerdman, Orbital angular momentum of light and the transformation of Laguerre-Gaussian laser mode. Phys. Rev. A **45**, 8185–8189 (1992)

29. V.V. Kotlyar, S.N. Khonina, V.A. Soifer, Light field decomposition in angular harmonics by means of diffractive optics. J. Mod. Opt. **45**, 1495–1506 (1998)
30. S.N. Khonina, V.V. Kotlyar, V.A. Soifer, P. Paakkonen, J. Turunen, Measuring the light field orbital angular momentum using DOE. Opt. Memory Neural Networks **10**, 241–255 (2001)
31. T. Kaiser, D. Flamm, S. Schroter, M. Duppare, Complete modal decomposition of optical fibers using CGH-based correlation filter. Opt. Express **17**, 9347–9356 (2009)
32. M.V. Vasnetsov, J.P. Torres, D.V. Petrov, L. Torner, Observation of the orbital angular momentum spectrum of a light beam. Opt. Lett. **28**, 2285–2287 (2003)
33. A. Forbes, A. Dudley, M. McLaren, Creation and detection of optical modes with spatial light modulators. Adv. Opt. Phot. **8**, 200 (2016)
34. A. D'Errico, R. D'Amelio, B. Piccirillo, F. Cardano, L. Marrucci, Measuring the complex orbital angular momentum spectrum and spatial mode decomposition of structured light beams. Optica **4**, 1350–1357 (2017)
35. S. Fu, Y. Zhai, J. Zhang, X. Liu, R. Song, H. Zhou, C. Gao, Universal orbital angular momentum spectrum analyzer for beams. PhotoniX **1**, 19 (2020)
36. A. Volyar, M. Bretsko, Ya. Akimova, Yu. Egorov, Measurement of the vortex and orbital angular momentum spectra with a single cylindrical lens. Appl. Opt. **58**, 5748–5755 (2019)
37. A. Volyar, M. Bretsko, Ya. Akimova, Yu. Egorov, Orbital angular momentum and informational entropy in perturbed vortex beams. Opt. Lett. **44**, 5687–5690 (2019)
38. J. Pinnell, V. Rodriguez-Fajardo, A. Forbes, Single-step shaping of the orbital angular momentum spectrum of light. Opt. Express **27**, 28009–28021 (2019)
39. L. Zhu, J. Wang, A review of multiple optical vortices generation: methods and applications. Fron. Optoelectr. **12**, 52–68 (2019)
40. Y. Yang, Q. Zhao, L. Liu, Y. Liu, C. Rosales-Guzman, C. Qiu, Manipulation of orbital angular momentum spectrum using pinhole plates. Phys. Rev. Appl. **12**, 064007 (2019)
41. J. Leach, M.J. Padgett, S.M. Barnett, S. Franke-Arnold, J. Courtial, Measuring the orbital angular momentum of a single photon. Phys. Rev. Lett. **88**, 257901 (2002)
42. C. Schulze, A. Dadley, D. Flamm, M. Duparre, A. Forbes, Measurements of the orbital angular momentum density of light by modal decomposition. New J. Phys. **15**, 073025 (2013)
43. P. Bierdz, M. Kwon, C. Roncaioli, H. Deng, High fidelity detection of the orbital angular momentum of light by time mapping. New J. Phys. **15**, 113062 (2013)
44. S. Li, P. Zhao, X. Feng, K. Cui, F. Liu, W. Zhang, Y. Huang, Measuring the orbital angular momentum spectrum with a single point detector. Opt. Lett. **43**, 4607–4610 (2018)
45. V.V. Kotlyar, A.A. Kovalev, A.P. Porfirev, Calculation of fractional orbital angular momentum of superpositions of optical vortices by intensity moments. Opt. Express **27**, 11236–11251 (2019)
46. A. Volyar, M. Bretsko, Ya. Akimova, Yu. Egorov, Vortex avalanche in the perturbed singular beams. J. Opt. Soc. Am. A **36**, 1064–1071 (2019)
47. E. Karimi, B. Piccirillo, E. Nagali, L. Marucci, E. Santamato, Efficient generation and sorting of orbital angular momentum eigenmodes of light by thermally tuned q-plates. Appl. Phys. Lett. **94**, 231124 (2009)
48. G.C.G. Berghout, M.P.J. Lavery, J. Cortial, M.W. Beijersbergen, M.J. Padgett, Efficient sorting of orbital angular momenyum states of light. Phys. Rev. Lett. **105**, 153601 (2010)
49. M. Mirhosseini, M. Malik, Z. Shi, R.W. Boyd, Efficient separation of the orbital angular momentum eigenstates of light. Nat. Commun.Commun. **4**, 1–6 (2013)
50. H. Zucker, Optical resonators with variable reflectivity mirrors. Bell Syst. Techn. Journ. **49**, 2349–2376 (1970)
51. F. Gori, G. Guattary, C. Padovani, Bessel-Gauss beams. Opt. Commun.Commun. **64**, 491–495 (1987)
52. V.V. Kotlyar, R.V. Skidanov, S.N. Khonina, V.A. Soifer, Hypergeometric modes. Opt. Lett. **32**, 742–744 (2007)
53. E. Karimi, G. Zito, B. Piccirillo, L. Marrucci, E. Santamato, Hypergeometric-Gaussian modes. Opt. Lett. **32**, 3053–3055 (2007)
54. M.A. Bandres J.C. Gutierrez-Vega, Circular beams. Opt. Lett. **33**, 177–179 (2008)

55. I.V. Basisty, V.Yu. Bazhenov, M.S. Soskin, M.V. Vasnetsov, Optics of light beams with screw dislocations. Opt. Commun. **103**, 422–428 (1993)
56. V.V. Kotlyar, A.A. Kovalev, A.P. Porfirev, Elliptic Gaussian optical vortices. Phys. Rev. A **95**, 053805 (2017)
57. V.V. Kotlyar, A.A. Kovalev, V.A. Soifer, Asymmetric Bessel modes. Opt. Lett. **39**, 2395–2398 (2014)
58. J. Courtial, K. Dholakia, L. Allen, M.J. Padgett, Gaussian beams with very high orbital angular momentum. Opt. Commun.Commun. **144**, 210–213 (1997)
59. J. Serna, J. Movilla, Orbital angular momentum of partially coherent beams. Opt. Lett. **26**, 405–407 (2001)
60. S.N. Khonina, V.V. Kotlyar, V.A. Soifer, P. Paakkonen, J. Simonen, J. Turunen, An analysis of the angular momentum of a light field in terms of angular harmonics. J. Mod. Opt. **48**, 1543–1557 (2001)
61. I. Martinez-Castellanos, J. Gutiérrez-Vega, Shaping optical beams with non-integer orbital-angular momentum: a generalized differential operator approach. Opt. Lett. **40**, 1764–1767 (2015)
62. Y. Yang, L. Wu, Y. Liu, D. Xie, Z. Jin, J. Li, G. Hu, C. Qiu, Deuterogenic plasmonic vortices. Nano Lett. **20**, 6774–6779 (2020)
63. D.V. Petrov, Vortex-edge dislocation interaction in a linear medium. Opt. Commun.Commun. **188**, 307–312 (2001)
64. D.V. Petrov, Splitting of an edge dislocation by an optical vortex. Opt. Quant. Electr. **34**, 759–773 (2002)
65. V.V. Kotlyar, A.A. Almazov, S.N. Khonina, V.A. Soifer, H. Elfstrom, J. Turunen, Generation of phase singularity through diffracting a plane or Gaussian beam by a spiral phase plate. J. Opt. Soc. Am. A **22**, 849–861 (2005)
66. M.V. Berry, Optical currents. J. Opt. A Pure Appl. Opt. **11**, 094001 (2009)
67. V.V. Kotlyar, A.A. Kovalev, Optical vortex beams with a symmetric and almost symmetric OAM spectrum.J. Opt. Soc. Am. A. **38**, 1276–1283 (2021)
68. D. Hebri, S. Rasouli, Combined half-integer Bessel-like beams: A set of solutions of the wave equation. Phys. Rev. A **98**, 043826 (2018)

Conclusion

This book contains the authors' results on helical beams obtained in 2023-2024. We will briefly list these results.

We have derived novel exact solutions of the paraxial propagation equation. Relationships have also been deduced for complex amplitudes that describe the propagation of generalized asymmetric HG and LG beams. The derived relationships contain parameter sets, with an amplitude of a generalized asymmetric LG beam described by two complex parameters, namely, the scale parameter and the asymmetry degree parameter. The amplitude of a generalized asymmetric HG beam contains four parameters, a pair for each Cartesian axis. With the original generalized elegant HG and LG beams not retaining their intensity pattern upon free-space propagation, the derivative generalized asymmetric HG and LG beams also do not retain their intensity pattern in the course of their propagation evolution. The generalized LG beams are not orthogonal with regard to the radial indices, meaning that the asymmetric LG beams are also non-orthogonal with regard to the radial indices. For the generalized LG beams, we have proved that their topological charge is independent on the asymmetry and equals the upper index of the Laguerre polynomial. We have also found that the asymmetric LG beam is anomalously rotated upon propagation in free space, i.e. it is rotated clockwise when the vortex factor has a positive topological charge.

We derived analytical expressions for 24 IG modes with the indices $p = 3, 4, 5, 6$ via 2, 3 or 4 LG modes (more exactly, via their real or imaginary parts) and HG modes The expansion coefficients of the IG modes by the LG and HG modes are expressed via the ellipticity parameter ε. Using the expressions obtained, one can immediately derive expressions for the IG modes when the parameter ε either equals zero or tends to plus (or minus) infinity. Explicit dependence of the IG modes on the ellipticity parameter allows controlling their intensity by varying this parameter. We have derived the symmetry properties for the even and odd IG modes. In particular, we demonstrated how these modes are interrelated when the ellipticity parameter changes its sign. The obtained representations of the IG modes via the HG and LG modes are not only convenient from the theoretical point of view, since they reveal the properties of these beams without simulation, but also from the practical point of view, since they allow rather easily programming complex amplitudes of the IG

V. V. Kotlyar et al., *Helical Laser Beams*, https://doi.org/10.1007/978-3-031-92217-6

modes for their numerical simulation or generation by using a spatial light modulator. Explicit analytical formulae for complex amplitudes of hIG modes at $p = 2 - 5$ have been derived. Using those equations, one can easily deduce expressions for hIG modes at the limiting cases, $\varepsilon = 0$ and $\varepsilon = \pm\infty$, which are in compliance with the general relationships. For the hIG modes studied, we have also deduced explicit relationships for the OAM as a function of the parameter ε and in the limiting cases. In the general case, the ε-dependence of the OAM has been shown to be an even function. It has been shown that at $\varepsilon = 0$, the TC of hIG modes equals that of a LG mode, i.e. the index q, being equal to the index p at $\varepsilon = +\infty$.

We have introduced a four-parameter family of vortex beams overlapping well-known LG beams. The new beams represent a product of two different LG beams with the same radius of waist and can also be termed as a product of LG (pLG) beams. If in a pLG beam both Laguerre polynomials have the same indices, such a beam may be referred to as a 'squared' LG beam, or $(LG)^2$ beam. The proposed pLG beams have been expressed as superposition of a finite sum of conventional LG beams. For the $(LG)^2$ beams, an explicit Fourier transform has been derived. A particular case of pLG beams whose Laguerre polynomials are described by specially tailored azimuthal indices n-m and $n + m$ has been analyzed and their Fourier transform has been deduced in an explicit form.

We obtain and investigate a new one-parametric family of Bessel-Bessel-Gaussian beams (BBG beams). We derive their complex amplitude as a product of the Gaussian function and two nth-order Bessel functions with a similar dependence of their arguments on the cylindrical coordinates. It is shown that changing the positive beam parameter allows controlling the beam topological charge. We also demonstrate anomalously fast transverse rotation of the beam intensity pattern upon propagation in the near field of diffraction. At a distance from the waist much shorter than the Rayleigh length, the intensity is already rotated by almost 45°.

Preface

In scalar optics, light is described by a complex amplitude, a complex function of three Cartesian coordinates. This function must be a solution to the scalar paraxial Helmholtz equation, which is equivalent to the Schrödinger equation in quantum mechanics. There are not many known exact analytical solutions of this equation in the form of special functions, only a few dozen. Each such solution can be associated with a certain laser beam, for example, a Bessel, Laguerre-Gaussian or Hermite-Gaussian beam. Each such analytical solution of the Helmholtz equation allows one to fully describe all the features of the light beam before modeling. Find the intensity distribution at any distance from the waist, phase distribution, total beam power and its other characteristics. Therefore, the search for new analytical solutions describing new laser beams, including helical (vortex) beams, which possess orbital angular momentum and topological charge, is relevant. This book describes new helical beams that the authors obtained in 2023–2024. These are generalized asymmetric Laguerre-Gaussian and Hermite-Gaussian beams, double and square Bessel-Gaussian and Laguerre-Gaussian beams, and several types of Bessel-Bessel-Gaussian beams. The book also discusses helical Ince Gaussian beams, which are presented as series expansions in Hermite-Gaussian modes, and analytical expressions for the orbital angular momentum are obtained for them. Each such new analytical solution of the Helmholtz paraxial equation is a significant contribution to optics.

The book will be of interest to a wide range of scientists, engineers working in the field of optics, photonics, laser physics, opto-information technologies, and optical instrumentation. It can also be useful for bachelors and masters in the specialties applied mathematics and physics, applied mathematics and informatics, optics and graduate students specializing in these areas.

Samara, Russia

Victor V. Kotlyar
Eugeny G. Abramochkin
Alexey A. Kovalev

The manufacturer's authorised representative in the EU is Springer Nature Customer Service Centre GmbH, Europaplatz 3, 69115 Heidelberg, Germany. If you have any concerns regarding our products, please contact ProductSafety@springernature.com

Printed and bound by CPI Group (UK) Ltd, Croydon, CR0 4YY
30/09/2026
02225604-0017